EDI and X.400 using P$_{edi}$

EDI and X.400 using P$_{edi}$
The Guide for Implementors and Users

Richard Hill
Hewlett-Packard

Technology  Appraisals

Technology Appraisals Ltd.

Technology Appraisals specialises in top quality professional education and information for users and suppliers of computer communications. Products include seminars, publications and OSN - The Open Systems Newsletter. For further information please write or call: Technology Appraisals Ltd, Grove House, 551 London Road, Isleworth, TW7 4DS, United Kingdom; Telephone: +44-81-744 1155, Fax: +44-81-744 1149, Dialcom/Gold: 87:SQQ255.

BRITISH LIBRARY CATALOGUING-IN-PUBLICATION DATA

Hill, Richard
    EDI and X.400 using P_{edi}
    :The Guide for Implementors and Users
    1.   Data processing
    I.   Title
    004

ISBN   1 871802 05 9

Printed and bound in Great Britain
at The Camelot Press, Trowbridge, Wiltshire.

To the memory of Johan Lundberg, who contributed so much.

To Ted Myer, who made P_{edi} a reality.

To Charlie Combes, Dick Jesmajian, Carl-Uno Manros, Dave McKnight, John Pilkington and John Ross, who did much of the writing and editing.

To Tim Bishop, who made it fun, and Chris Brook, who persevered.

To the rest of the group, who made it possible.

Acknowledgements

I would like to thank my employer, Hewlett-Packard, who has generously supported both the work of the CCITT Associate Rapporteur Group for EDI and X.400 and the production of this book.

This book has been composed, edited and camera-ready masters printed using Hewlett-Packard hardware, Microsoft Word 5.0, and the HP Graphics Gallery graphical package.

Disclaimer

The views and opinions expressed in this book are those of the author, and do not necessarily represent those of the Hewlett-Packard company.

This book represents the author's understanding of the Recommendations F.435 and X.435, on the basis of the text available at the time of publication. Implementations should <u>not</u> be based on this book: they should be based on the official CCITT Recommendations which will be published in the near future.

This book does not represent the views of any private or public organization, nor the views of any standard-making body.

The author and the publishers decline responsibility for any errors of commission or omission that may be contained in this book.

Table of contents

Examples

Figures

Tables

Foreword

This book marks the merge point between two revolutions, one in communication, the other in business practice. Both revolutions trace their roots back to relatively obscure beginnings more than two decades ago.

The first began with early experiments in computer based messaging. It has since spawned a small industry in electronic mail services and systems. Along the way, this revolution gave rise to the CCITT X.400 Recommendations for Message Handling Systems, a set of international standards that have become a driving force in the messaging industry. The X.400 Recommendations provide enabling specifications for the construction, now underway, of a global messaging network whose scope and utility will eventually rival those of the telephone and postal systems.

Much of X.400's promise arises from the designed-in capability of an X.400 conformant network to simultaneously support many applications each generating a separate kind of traffic, for example document distribution, funds transfer, voice text and facsimile messaging. However, as of 1988, when the second generation of the X.400 Recommendations were ratified, only one application had been standardized - the exchange between human users of informal, memo-style messages.

The second revolution began with the realization, initially in North America, that commercial growth threatened to bury many industries in the paperwork associated with postal exchange of trade documents such as purchase orders, shipping notices and invoices. It became clear that if the exchange could be accomplished electronically, then problems of delay, cost and errors would be reduced or eliminated. The term Electronic Data Interchange (EDI) was eventually coined to describe the resulting worldwide movement towards paperless electronic exchange of business transaction documents.

Over the years, EDI received much attention from the standards community, initially in North America and more recently through the UN supported EDIFACT organization. The most significant achievement of these groups has been to develop an encoding system for electronically representing trade documents, and to then apply that system so as to produce encoding specifications for many specific document types. The problem of communicating such documents has received less attention from these groups. EDIFACT, for one, made clear that the communication community was expected to develop a standard solution for the actual transfer of EDIFACT-specified documents.

Thus, by 1988, there had arisen a synergistic situation: a capable but under-exploited communication standard on the one hand, and, on the other, a set of document specifications seeking the support of a communication standard.

Growing recognition of this situation led CCITT in 1988 to commission a project to bring together EDI and X.400 in the form of a standardized X.400 solution for EDI transfer. The result of that project, two new CCITT Recommendations, F.435 and X.435, describe the service and technical sides of that solution, and form the subject of this book.

This book complements and extends the material in the standards documents in two important ways. First, the group commissioned by CCITT was joined by a set of participants unusually rich in its diversity, with representatives of the user, EDI standards, and provider communities. This gave rise to lively debates in which every part of the prospective solution was questioned, dissected and repeatedly reassembled until clearly able to meet the intended need, whose validity itself was held up to constant question. The thought process behind these debates is, of course, missing from the standards documents, which describe the shape of the solution but not the logic that yielded that shape. This book supplies the missing insight into why things are the way they are in EDI and X.400.

Second, the standards documents, again deliberately, steer clear of guidance on how the X.400 solution should be implemented or how it might fit into the practice of EDI. Again, this book provides the missing insight.

The book's author was also one of the most productive and stimulating members of the group. Coming at the problem primarily from the user side he constantly challenged the presumptions of the communications community and came forth with new and imaginative variants in their technology. He also served as editor of X.435, taking on the full challenge of that complex document under crisis conditions engendered by the tragic and sudden death of its original editor. It is safe to say that the project could not possibly have been completed on time had Richard Hill not stepped up to the task.

Theodore H. Myer
CCITT Associate Rapporteur for EDI and X.400
Washington, D.C., June, 1990

Preface

This book has been written for everyone who has a basic working knowledge of both EDI and the X.400 series of Recommendations, and who would like to understand in detail how the new CCITT X.400 standard for EDI (X.435, which defines the P_{edi} protocol) can be used and implemented.

The major objective of the book is to guide the reader through the complexities of X.435, by presenting the content of X.435 in a modular, logical fashion, and by providing figures, examples and explanatory text that are not present in the standards themselves.

This book was written based on the final drafts of F.435 and X.435 available in June 1990. These final drafts were presented to CCITT, and are expected to be approved by the CCITT plenary meetings of Study Group I for F.435 and Study Group VII for X.435 in November 1990. Minor changes to the text of the Recommendations are possible before their final approval and publication by CCITT; these changes, if any, are not reflected in this book. Considering the low likelihood of significant changes, and the high interest in the new protocol, the publishers and I felt that it was in the best interests of the public to publish this book before final approval of the CCITT Recommendations.

Actual implementations should of course be based on the final approved CCITT Recommendations, and not on the text of this book.

ISO/IEC have indicated their intention to take over the text of X.435, and approve it as an International Standard during 1991. The ISO text will be identical in substance to the CCITT text.

I would like to thank all the people in CCITT, ISO and EDIFICE who have contributed to the writing of F.435 and X.435. First of all Johan Lundberg, whose passing still saddens us.

Special thanks go to Ted Myer, the CCITT Associate Rapporteur for EDI and X.400, without whom P_{edi} would not exist today, to Charlie Combes, Dick Jesmajian, Carl-Uno Manros, Dave McKnight, John Pilkington and John Ross, who did much of the writing and editing, to the other members of the group, whose skills and knowledge were invaluable, to my colleagues in EDIFICE, who helped to insure that the EDI users' point of view was well represented in the group.

I would also like to thank David Hitchcock of Technology Appraisals for his commitment to this book and for his editorial support, and, last but not least, my employer, the Hewlett-Packard company, who has generously supported both the work of the CCITT Associate Rapporteur group for EDI and X.400 and the writing of this book.

Richard Hill
Coinsins, Switzerland, July 1990

1 Introduction

It has been clear for some time that the basic nature of the X.400 data transmission protocol is well suited to the needs of Electronic Data Interchage (EDI) users. In fact, many believe that X.400 can provide the essential services commonly provided today by EDI-specialist clearing house services (commonly called Value Added Network services, or VANs).

Use of X.400 for EDI has been impeded by two factors:

1. Lack of inexpensive, programmatically callable software that implements User Agents (UAs) and Message Transfer Agents (MTAs).

2. Lack of an X.400 protocol specifically designed for EDI.

The first issue is being addressed by software and hardware manufacturers, who are producing new, less expensive hardware and software, and who will soon be producing programmatically callable UAs and MTAs based on the API Association standards.

The second issue has been addressed by the CCITT, which has created the Associate Rapporteur Group for EDI and X.400, whose charter has been to define extensions to X.400 needed to accommodate EDI user needs.

The Group, under the leadership of Ted Myer, started work in August 1988, and completed its work in June 1990. The completed work consists of the Recommendations F.435 and X.435, which define the P_{edi} protocol. The Postscript to this book provides some insight into the design principles that helped to guide the development of P_{edi}.

The P_{edi} protocol contains several features that are unique to it, and that are not found in other X.400 protocols, nor in proprietary data transmission protocols commonly used today for EDI. Among the features worth mentioning are:

- forwarding of responsibility

- end-to-end acknowledgments

- security

As will be shown in Chapter 2, the P_{edi} protocol can be used to satisfy a wide range of EDI user needs. Because of its generality, the protocol is rich and rather complex.

This, together with the fact that, as normal for international standards and CCITT Recommendations, the documents F.435 and X.435 do not contain much tutorial or explanatory material, means that it may be quite difficult for either a user or a software producer to understand what parts of the protocol should be implemented and used for a particular application.

The objective of this book is to help both EDI users and X.400 software developers to understand better how the several features of the P_{edi} protocol can be used in any particular case. The book includes much tutorial and informative material that was presented and discussed within the Associate Rapporteur Group, but that was not considered suitable for inclusion in an international standard.

A large part of the value-added of this book consists in pointing the reader to the relevant clauses of the Recommendations F.435 and X.435. If one knows where to look, the Recommendations are clear. The difficulty is that it is not always obvious where to look to find the answer to a specific question.

This book will help the reader to find his way through the Recommendations, thereby reducing the number of times that the Recommendations are scanned without understanding, merely in order to find out which parts need to be read and in which order.

X.400 implementors, Value Added Network providers and EDI users familiar with X.400 will wish to read all Chapters of this book (except for section 1.1).

EDI users who are not familiar with X.400 should read section 1.1, and they may then wish to read the following portions of this book, in order to gain an overall understanding of how P_{edi} can be used to transmit EDI Interchanges:

- Chapter 2

- Sections 3.1, 3.2 and 3.3

- First page of Chapter 4

- Sections 5.1, 5.2 and 5.3

- Section 6.1

- Chapter 7

- Chapter 8

- Chapter 10

While this book is a tutorial on the P_{edi} protocol, it is not a tutorial on X.400 itself. Subsequent chapters assume that the reader is familiar with the basic concepts of X.400 (specifically the 1988 version of X.400). Thus, acronyms such as MTA, UA, P2, IPM and standard X.400 terms such as ORName will be used freely, with no explanation. Section 1.1 provides a very basic overview of the key concepts of X.400; it can be skipped by readers who are familiar with X.400.

The reader is also assumed to be familiar with the basic concepts of EDI, and standard EDI terms such as *segment* will be used with no explanation. Section 1.2 provides a very basic overview of the key concepts of EDI; it can be skipped by readers who are familiar with EDI.

The glossary at the end of the book does contain a definition for all acronyms used, and for the key X.400 and EDI terms used.

The index includes entries for all references that are made to X.400 standards within the book. References to F.435 and X.435 are indexed by clause number.

NOTE: One possible hindrance to the understanding of international standards is their use of common words in a specific technical sense. For example, the word *message* has a very specific definition in the X.400 Recommendations, and quite a different very specific meaning in the EDIFACT standard.

In this book, all words that are neither italicized nor capitalized are used with their normal common meaning. That is, if you do not understand a word used in this book, you may look it up in a normal English language dictionary (a few words, such as *application*, are used in a computer-specific sense, and would have to be looked up in a dictionary of computer-specific English language usage).

When there is a need to use a term defined in an
international standard, and the term could be confused with
a common word, the term will either be italicized (for
example, *delivery*), or prefixed with an acronym and
capitalized (for example, EDI Responsibility), or both.

1.1   Rudiments of X.400

The X.400 series of Recommendations provides an international standard
for structuring and transmitting electronic mail messages. Analogies to well-
known physical systems can help to understand X.400.

If we compare X.400 to the telephone system, we can say that X.400 is
equivalent to the standards that govern the signals transmitted along the
phone lines: these signals are used for switching (selecting telephone
numbers), and for transmitting data. Telephone standards do not specify the
shape and functions of the physical phones used by the end-users; similarly,
X.400 standards do not specify the structure and functions of the electronic
mail programs accessed by end-users. X.400 specifies the structure of the
data that one electronic mail program sends to another electronic mail
program.

Regulated providers (formerly called PTTs in Europe), provide basic
carrier and switching services for telephones. The equivalent concept in
X.400 is the Administrative Management Domain (ADMD).

An organization that owns a certain number of private telephones will
normally own (or lease) and operate a Private Branch Exchange (PBX).
The equivalent concept in X.400 is the Private Management Domain
(PRMD).

If we compare X.400 to physical mail, we can say that X.400 is a series of
standards (rules) that would be analogous to standards governing the size
and shape of envelopes, the structure and content of addresses on envelopes
and the structure of what is contained in an envelope.

The ADMD corresponds to public mail services, the PRMD to in-house
mail rooms.

Two of the key objects defined in X.400 are the User Agent (UA) and
the Message Transfer Agent (MTA). Both are processes (implemented in
hardware and software) that are responsible for providing certain services.
The UA provides the services that allow users to send and receive messages.
The MTA provides the services that allow messages to be moved from one
site to another site. Figure 1.1 illustrates these concepts.

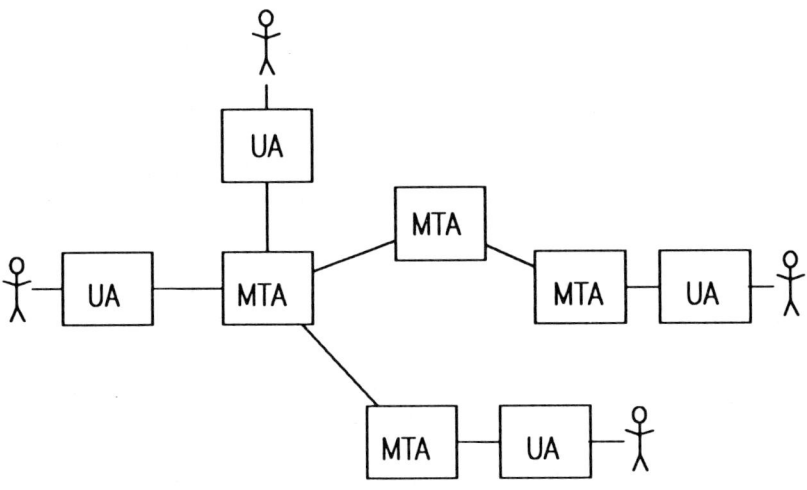

Figure 1.1 - The components of X.400

Note that, in the case of EDI, there are no human users, and the people shown in Figure 1.1 will be replaced by EDI processes (computer programs).

Using analogies to physical mail, we can explain the key concepts of X.400 as follows:

ADMD: A public message delivery service.

IPM: Inter-Personal Message. Same as P2, see below.

MTA: A computer program that actually moves *messages* from one site to another, but that is not capable of accepting *messages* from users or delivering *messages* to users (see UA). Similar in concept to the back-room operations of physical mail systems.

Notification: *Messages* that are used in order to confirm receipt of another *message*. Similar in concept to the return receipt of a registered letter.

ORName: Specific form of address required by X.400.

P1:   Protocol that defines the *envelope*.

P2:   Protocol that defines the structure of a memo.

PRMD: A private *message* delivery service.

UA: A computer program that accepts *messages* from users or delivers *messages* to users, but that does not actually move *messages* from one site to another (see MTA). Similar in concept to user-accessible post offices of physical mail systems.

X.400 can be an attractive data transmission mechanism for EDI data, because it provides the basic functions required by EDI users:

1. Universal connectivity: access to many other users, world-wide, using the same standard data transmission mechanism.

2. Store-and-forward: the user passes the message to X.400, who takes responsibility for delivering it, even if the recipient's computer is temporarily not available.

    That is, X.400 resembles physical mail more than the telephone system: you cannot transmit data by telephone unless the other party answers the phone (or has a machine that answers the phone).

    You can transmit a letter even if the other party is closed for vacations. The post office will hold the letter for a specified period of time, and allow deferred retrieval.

    X.400, like other store-and-forward networks, allows the sender to transmit messages without regard to whether or not the recipient is on-line and ready to receive them.

1.2   Rudiments of EDI

Electronic Data Interchange (EDI) can be defined as the computer-to-computer exchange of data related to commercial transactions using agreed upon formats and networks. EDI within a company is an old concept, and many large companies use EDI for essentially all intra-company business transactions (for example, intra-company invoices).

EDI with external trading partners is relatively new; its implementations are growing rapidly, and have already changed the way business is done in entire industries. In most cases EDI has improved the efficiency of the business process; in a few cases EDI has become a requirement for doing business.

There are two fundamental reasons to implement EDI: one is to increase profits by lowering the cost of the business process, the other is to increase market share by providing an improved service to customers.

The benefits of EDI include:

-    Faster transmission of data, resulting in shorter business cycle times.

-    More accurate transmission of data, and reduction in manual data-entry errors.

-    Reduced clerical overhead, since operations that were previously performed manually can be eliminated or automated with EDI.

The costs savings associated with EDI can be quite large for major companies, and many large companies are prepared to make significant investments in order to be able to reap the benefits of EDI.

1.2.1   Data encoding formats

A key component of EDI consists of the transfer of data according to standardized formats. The standardized formats are built up hierarchically, using definitions of the following:

1.   *Data elements*, which contain well-defined data, for example a <u>Product Number</u>.

2.   *Segments*, which consist of related *data elements* that are grouped together, for example <u>Name and Address</u>.

3.   *EDI Messages* (*Transaction Sets* in ANSI), which consist of *segments* that are grouped together to form a business transaction, for example a <u>Purchase Order</u>.

4.   *Interchanges*, which consist of *EDI Messages* that are grouped together for transmission, for example a batch of <u>Purchase Orders</u>.

Commonly used EDI standards are character-based, that is, all data are encoded in ASCII or equivalent character representations, and transmitted in this representation. Separator characters are used to delimit *data elements* and *segments*. Special *segments* are used to delimit *EDI messages* and *interchanges*.

The differences among the several commonly used standards (for example, EDIFACT, ANSI X12, TRADACOMS) consist primarily in differences among the separator characters, the definitions of *segments* and the definitions of *EDI messages*.

All commonly used EDI standards include the concept of a special *segment* that is used to start an *interchange*. This *segment* is commonly referred to as the *header segment*. The key *data elements* contained in this *segment* are common to most EDI standards, and it is these *data elements* which are of most interest to P_{edi}, because these *data elements* include some information that can be used for addressing, selective retrieval and routing.

We show below a very simple example of an EDIFACT Purchase Order.

```
UNB+UNOA:1+sender+recipient+date:time+reference number'
UNH+reference number+ORDERS:1'
BGM+105+purchase order number+date'
NAD+ST+++ship-to address'
NAD+SE+++supplier address'
CUX+currency code:OC'
UNS+D'
LIN+1++part number:VP++quantity+price'
DTM+002+requested delivery date'
UNS+S'
UNT+10+reference number from UNH'
UNZ+1+reference number from UNB'
```

When the data are transmitted, all blanks are removed, and the resulting character string is transmitted as one block.

1.2.2   EDI processing model

Agreement on a common standardized data format is not sufficient for EDI, because some method of transmitting data across networks must be used, in order to move the data from an application running on a computer in a particular site at one company to another application running on another computer in a particular site at another company.

Figure 1.2 illustrates one common model that is used to decompose the EDI process into manageable steps:

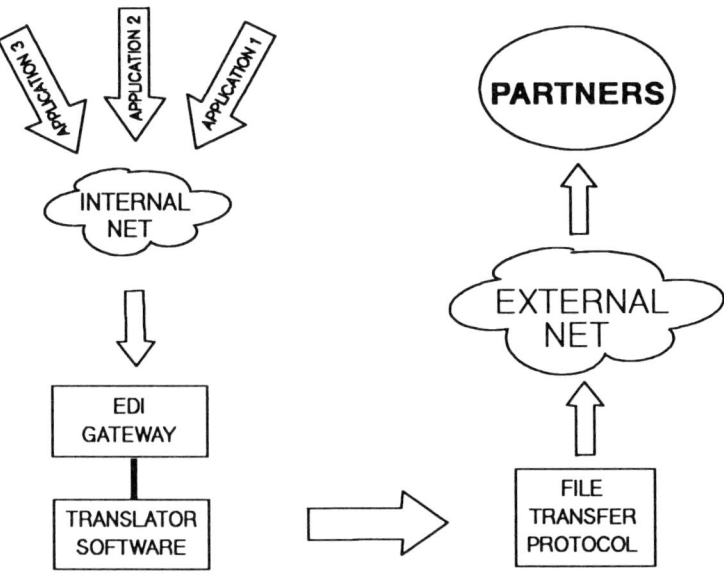

Figure 1.2 - EDI processing model

In a large company, many applications run at many sites. While it is possible to connect each application directly to trading partners, most large organizations find it convenient to use an internal network to transfer data to a limited number of corporate gateways.

These gateways typically convert data from internal formats to standard EDI formats, and then transfer the data to partners, using public or private external networks, and proprietary or standard file transfer protocols.

In this context, a file transfer protocol is a telecommunications protocol that ensures that the EDI Interchanges are correctly transmitted as indivisible blocks, and are not broken up or scrambled in any way.

While numerous private external networks and proprietary file transfer protocols exist, and are commonly used, there is clearly an advantage in using internationally standardized file transfer protocols, just as there is an advantage in using internationally standardized EDI data formats instead of proprietary EDI data formats.

Public network providers (PTTs) do not normally provide services based on proprietary protocols, but may provide services based on internationally standardized protocols.

P_{edi}, in conjunction with the other X.400 Recommendations, provides an internationally standardized protocol that is specifically designed to provide file transfer and networking for EDI.

Referring back to Figure 1.2, P_{edi} can be used to implement the internal network, the file transfer protocol and the external network portions of the EDI processing model. P_{edi} is one of the few telecommunications protocols that has such wide applicability to EDI.

1.3   Use of 1984 X.400 for EDI

Before discussing P_{edi} in detail, it is useful to review briefly how the 1984 version of X.400 has been used to implement EDI, and to show the basic structure of a P_{edi} *message*. P_{edi} is based on the 1988 version of X.400. Chapter 13 discusses interworking of P_{edi} and existing implementations based on the 1984 version of X.400.

An EDI Interchange is just a character string. If EDIFACT or TRADACOMS are used as EDI syntax standards, the EDI Interchange consists of ASCII characters. If ANSI X12 is used as the EDI syntax standard, the EDI Interchange may consist of EBCDIC characters.

There are two ways to use 1984 X.400 to transport an EDI Interchange:

-      the P2 approach, used in Europe

-      the P1/0 approach, used in the United States

1.3.1   P2 approach

In the P2 approach, the EDI Interchange is transported as the text part of an Inter Personal Message (IPM) *content*.

This approach is used in Europe, and has been standardized by the Commission of the European Communities.

The body of the IPM contains the EDI Interchange, coded as IA5String. Figure 1.3 illustrates this approach.

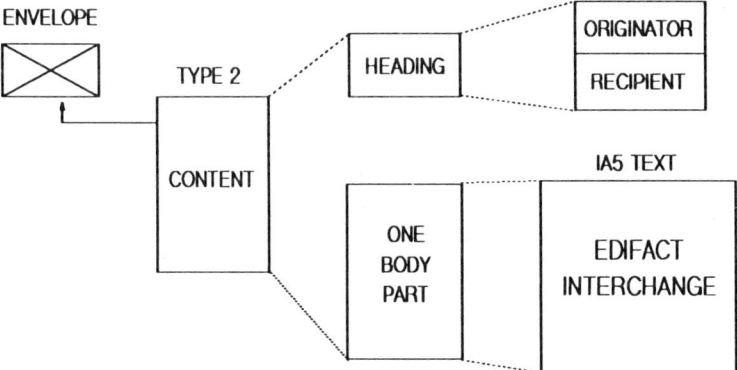

Figure 1.3 - P2 Approach

Converting the EDI Sender and Recipient in the EDI *header segment* (for example, the UNB *segment* in EDIFACT) to *originator* and *recipient* ORAddresses is a local matter. No rules are specified for how to do this.

Sections 13.1 and 13.2 discuss interworking P_{edi} and the P2 approach.

1.3.2 P1/0 approach

In the P1/0 approach, the EDI Interchange is transported as *content* with type "undefined", directly in the *envelope*.

This approach is used in the United States, and has been standardized by the National Institute for Technical Standards and by ANSI X12. Figure 1.4 illustrates this approach.

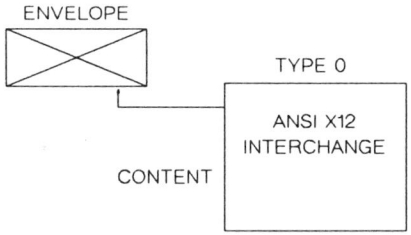

Figure 1.4 - P1/0 Approach

Sections 13.3 and 13.4 discuss interworking of P_{edi} and the P1/0 approach.

1.3.3   P$_{edi}$ overview

In contrast to the P2 and P1/0 approaches, P$_{edi}$ provides a *message* structure that is specifically designed for EDI Interchanges. The full details of this structure are explained in Chapter 3.

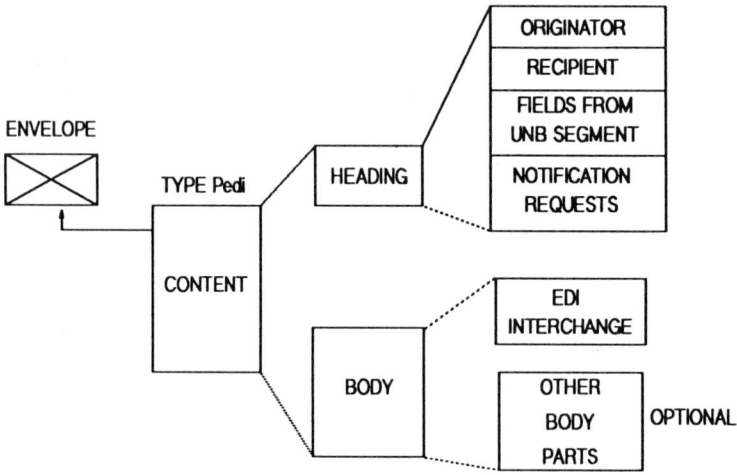

Figure 1.5 - P$_{edi}$ overview

Figure 1.5 illustrates the fundamental features of the P$_{edi}$ *message* structure: a *heading* that contains *fields* which contain data derived from the EDI *header segment* (for example, the UNB *segment* in EDIFACT).

2 Possible scenarios

There are many ways in which users implement EDI, and therefore there are many ways in which EDI users will wish to use X.400, if they choose X.400 as a data transmission mechanism for EDI.

The P_{edi} protocol includes features which will be useful for all of the following scenarios:

1. The EDI UA is associated with one particular EDI application.

2. The EDI UA is a corporate EDI gateway which accepts responsibility for all EDI Interchanges which it receives, and acknowledges receipt of the EDI Interchange if the sender requested acknowledgment.

3. The EDI UA is a corporate EDI gateway which does not accept responsibility for the EDI Interchanges it receives, and does not acknowledge receipt of the EDI Interchange. The EDI Interchange will be forwarded, and the final recipient UA will acknowledge receipt, if the sender requested acknowledgment.

4. The EDI UA is part of a clearing house (VAN) service.

5. The EDI UA is used by a small company, which forwards some of the EDI Interchanges for processing by a service bureau.

It should be noted that these scenarios are neither exhaustive nor mutually exclusive. Other scenarios are possible, and combinations of scenarios will be used in practice. For example, some corporations will combine scenarios 1 and 2, while others might combine 1 and 3.

Each of the scenarios is discussed in more detail below.

2.1   The UA serves one application

This is the simplest case, and the one which does not need all the features of the P_{edi} protocol. In this scenario, each application has its own ORName, and thus, implicitly, its own UA.

The application can be thought of as being tightly coupled to the UA, and the UA can be thought of as the telecommunications extension of the application.

The Figure 2.1 illustrates the scenario.

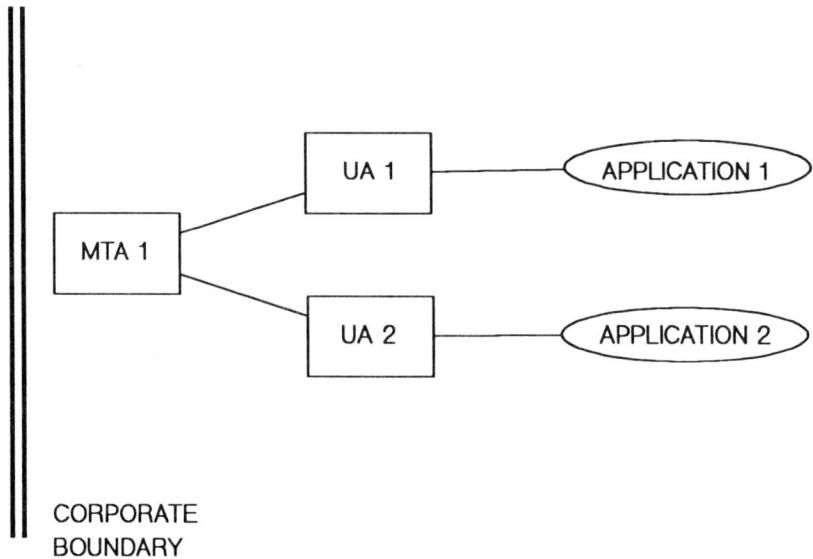

Figure 2.1 - UA serves one application

It should be noted that the corporate boundary could be to the right of MTA1, if the corporation does not wish to implement a PRMD. That is, MTA1 could be provided by an ADMD.

2.2   The UA is a corporate gateway that accepts responsibility

In this scenario, the corporation allows only one EDI UA (or a limited number of EDI UAs) to be visible to the outside world. The central gateway UA accepts responsibility for all EDI Interchanges that are received, and acknowledges receipt of the EDI Interchange, if acknowledgments were requested.

This is analogous to the operation of a central corporate mail room for physical mail. The mailroom staff signs for registered letters, and then forwards them within the corporation. The corporation is held to have received a letter when it has been received by the mail room.

Figure 2.2 illustrates the scenario.

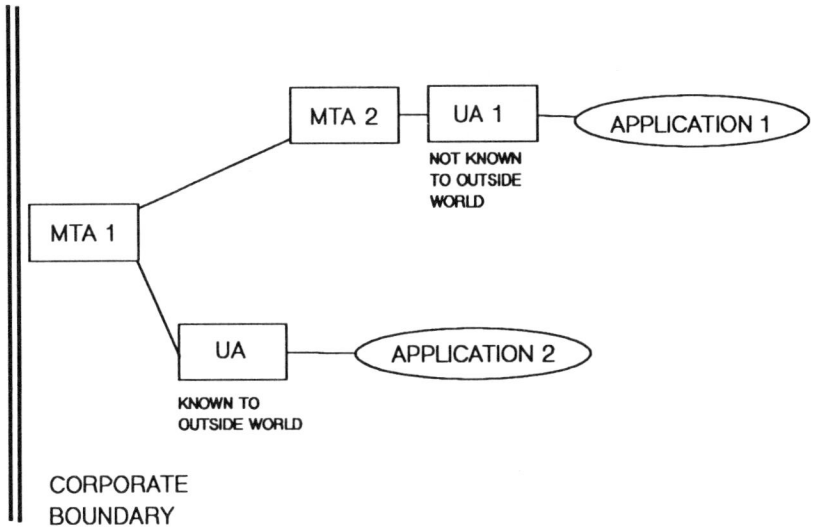

Figure 2.2 - Gateway UA accepts responsibility

It should be noted that the corporate boundary could be to the right of MTA1. That is, MTA1 could be provided by an ADMD.

Users outside the corporation do not address EDI Interchanges to UA1. UA1's address is only used within the corporation. That is, all EDI Interchanges originating outside the corporation are sent to the central gateway UA (UA in Figure 2.2), which passes them on to the applications. Similarly, applications within the corporation do not send EDI Interchanges directly to the outside world; they send the EDI Interchange to the central gateway UA, who then passes it on to the intended external recipient.

The corporation has the choice of using X.400 for transmitting EDI Interchanges from the central gateway UA to the application, or of using any other data transmission protocol, or of using any combination of X.400 and other protocols. This is illustrated in Figure 2.2 by the direct link of APPLICATION2 to the central gateway UA, and the use of X.400 to link APPLICATION1 to the central gateway UA.

That is, an EDI Interchange meant for APPLICATION1 will be sent to the central gateway UA, who will forward it to UA1, via MTA1 and MTA2, while an EDI Interchange meant for APPLICATION2 will be sent to the central gateway UA, who will directly pass it to APPLICATION2 (possibly using a proprietary data transmission protocol).

If X.400 is used for internal transmission, then the corporation must build a PRMD, illustrated by MTA2 in Figure 2.2.

A corporation can combine scenario 1 with scenario 2 by defining application-specific UAs and locating all the UAs on one corporate gateway computer. For example, if X.400 is not used for internal data transmission, so that MTA2 of Figure 2.2 does not exist, MTA1, UA1 and UA2 could all be located on a central gateway computer, while the applications are decentralized and run on other computers. In this case the central gateway UA illustrated in Figure 2.2 would be decomposed into UA1 and UA2, as illustrated in Figure 2.1.

2.3   The UA is a corporate gateway that forwards responsibility

In this scenario, the corporation allows only one EDI UA (or a limited number of EDI UAs) to be visible to the outside world. The central gateway UA does not accept responsibility for all EDI Interchanges that are received, and does not acknowledge receipt of the EDI Interchange if an acknowledgment was requested. The responsibility for acknowledging receipt is forwarded together with the EDI Interchange.

This is analogous to the operation of a poste restante or post office box (P.O. Box) service for physical mail. The post office staff does not sign for registered letters; the recipient himself (or his authorized delegate) must sign for the letter.

Figure 2.3 illustrates the scenario.

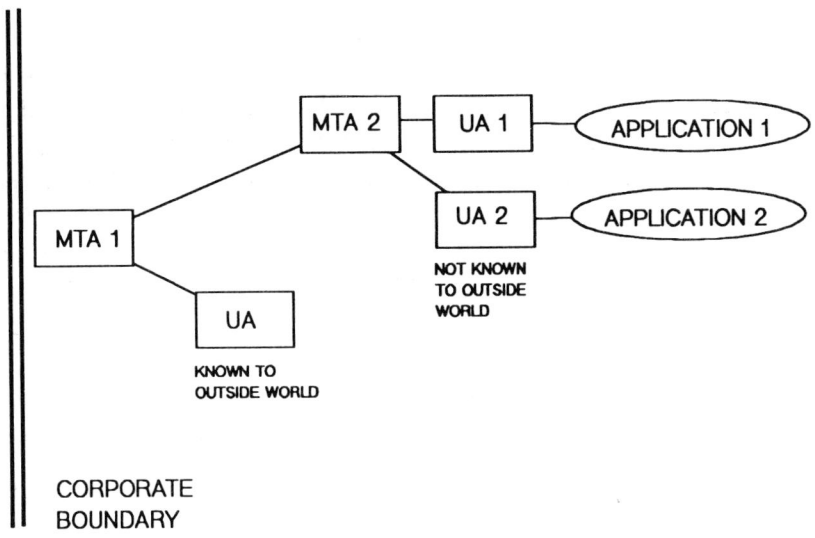

Figure 2.3 - Gateway does not accept responsibility

It should be noted that the corporate boundary could be to the right of MTA1. That is, MTA1 could be provided by an ADMD.

In this scenario, the corporation <u>must</u> use X.400 for transmitting EDI Interchanges from the central gateway UA to the application. The MTA2 illustrated in Figure 2.3 could be part of a PRMD, or it could be provided by an ADMD. The obligation to use X.400 for internal data transmission is implied by the fact that F.435 and X.435 impose rules for creating and sending *EDI Notifications*, and, in this scenario, these rules can only be satisfied if X.400 is used for internal data transmission.

A corporation can combine scenario 1 with scenario 3 by defining application-specific UAs and locating all the UAs on one corporate gateway computer. In this case, the central gateway UA illustrated in Figure 2.3 would be decomposed into UA1G and UA2G. The ORNames of UA1G and UA2G would be given to the outside world, and these UAs would forward EDI Interchanges to UA1 and UA2, whose ORNames would not be given to the outside world.

2.4   The UA is a clearing house

In this scenario, a private or public organization provides a clearing house service. The organization provides a collection and dispatching service for EDI Interchanges, on behalf of other organizations who are the actual EDI users. This type of service is commonly called a Value Added Network service (VAN). VAN providers often offer services in addition to the basic clearing house service described above (for example, syntax conversion, conversion to proprietary telecommunications protocols, audit trails, etc.).

Depending on its business charter, and on contractual agreements with its users, the clearing house may or may not accept responsibility for the EDI Interchange, it may or may not send acknowledgments for EDI Interchanges received, and it may or may not use data transmission protocols other than X.400.

This is analogous to the use of a service organization address for physical mail (lawyer's address, bank address). The service organization may or may not have authorization to sign for registered letters.

Figure 2.4 illustrates the scenario.

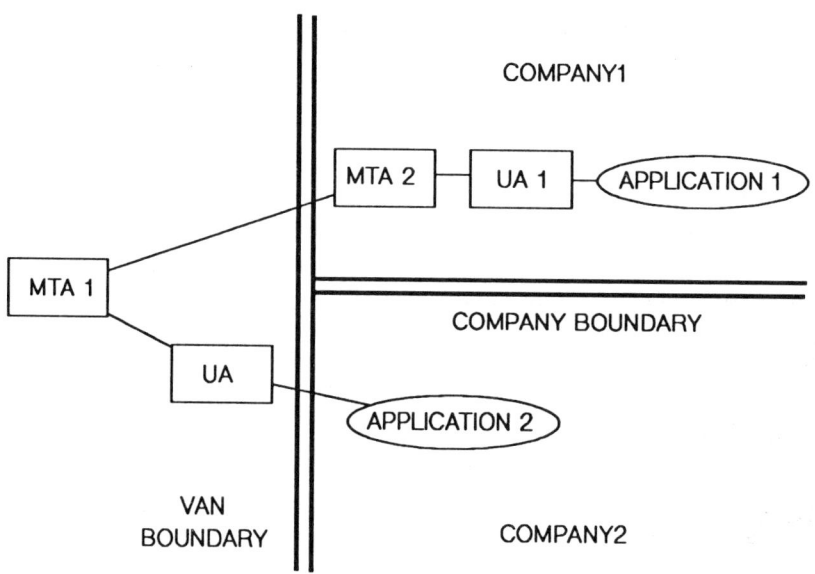

Figure 2.4 - VAN service

APPLICATION2 in Figure 2.4 belongs to a company (COMPANY2) that has not implemented X.400, and uses a UA provided by the VAN.

APPLICATION1 belongs to a different company (COMPANY1), that has implemented a PRMD, shown as MTA2 in Figure 2.4.

EDI Interchanges sent (by anyone) to COMPANY1 via the VAN can take two quite different paths. They could pass directly from MTA1 to MTA2, or they could pass from MTA1 to the VAN's UA, back to MTA1, and then to MTA2.

The path taken will depend on the way in which the VAN has implemented the P_{edi} protocol (in particular, on whether selective routing and forwarding of EDI Interchanges is done by the MTA operated by the VAN, or by the UA operated by the VAN). As we shall see later in the book, the P_{edi} protocol is sufficiently general to allow for different ways of providing clearing house services.

MTA2 could be part of a PRMD provided by COMPANY1, or it could be part of an ADMD, or it could be provided by the VAN itself (in which case MTA2 would not be needed: it would be identical to MTA1). MTA2 and/or UA1 could also be provided by a second VAN, not illustrated in Figure 2.4.

2.5   The UA is a small company gateway

In this scenario, a small company operates an EDI UA, and certain applications. Operation of other applications is delegated to an outside service bureau. The company's EDI UA accepts responsibility for certain EDI Interchanges, and does not accept responsibility for other EDI Interchanges.

For example, the company EDI UA could accept responsibility for all purchase orders (APPLICATION2), while it could delegate responsibility (to an accounting service) for all tax invoices (APPLICATION1).

Figure 2.5 illustrates the scenario.

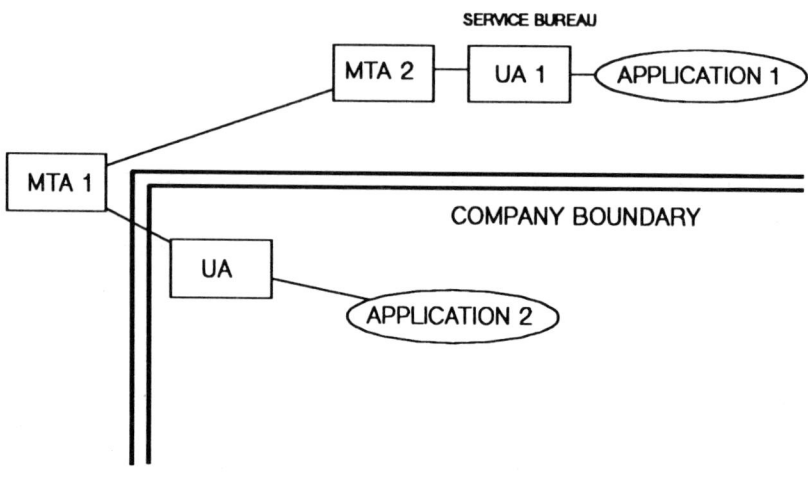

Figure 2.5 - Small company scenario

MTA1 could be provided by an ADMD, or by an authorized PRMD (VAN or clearing house).

3 Structure of an EDI Message

The heart of the P_{edi} protocol is a new X.400 *content type* specifically defined for EDI. This *content type* is called the *EDI Message type*.

The EDI Message (EDIM) is an X.400 *message* specifically designed to carry an EDI Interchange, just as the IPM *message* was designed to carry inter-personal messages (memos).

Optionally, an EDIM can carry other data related to the EDI Interchange (for example, a drawing related to an EDI Purchase Order). The EDIM is structured in such a way that the EDI UA can perform certain operations that are useful for EDI users, for example, forwarding and sending acknowledgments. These operations will be explained in detail in Chapters 5 and 6.

Before we try to understand the structure of an EDIM, we should understand how it fits in the X.400 world and how it relates to the EDI world.

Clauses 6 and 7 of F.435 specify how an EDIM fits into the X.400 world. Figure 6 of F.435 shows schematically how an EDI Interchange maps onto an EDIM; the essential features of this figure are captured in Figure 3.2 of section 3.1.

X.435 defines EDI-specific *user agents* (UAs) and *message stores* (MSs). The *message transfer service* (MTA) used by P_{edi} is the same MTA used by IPM. *Message stores* are optional.

Figure 3.1 shows the relationship between these concepts.

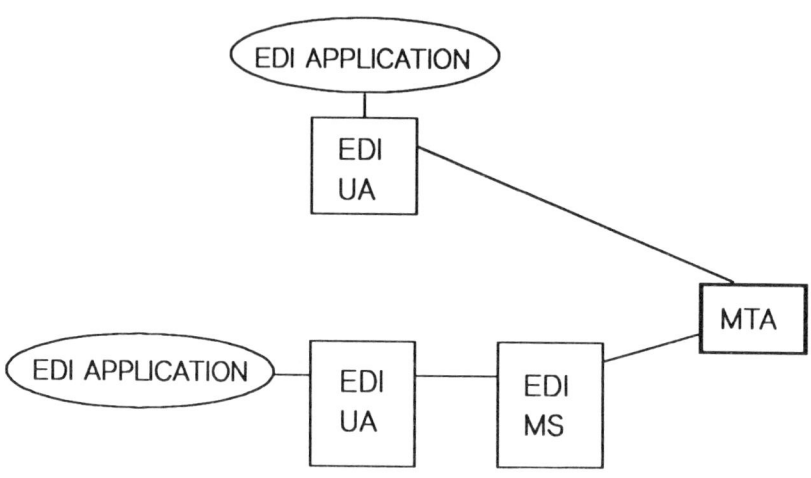

Figure 3.1 - The components of P$_{edi}$

The term *EDI Messaging User*, or *user* in F.435 and X.435 does <u>not</u> refer to a human user. It refers to the EDI process, that is to the computer systems that implement EDI applications.

The boundary between the EDI UA and the EDI Messaging User is not defined precisely in the X.400 standards. Considerable freedom is left for implementors to decide whether certain functions should be implemented in the EDI UA or in another process that is tightly coupled to the EDI UA. Sections 15.4 and 15.5 expand somewhat on this topic.

3.1   The EDIM

The EDIM itself is defined in clause 8 of X.435. The structure of an EDIM is illustrated in Figure 1 of clause 8 of X.435. The simplified Figure 3.2 shows the key features.

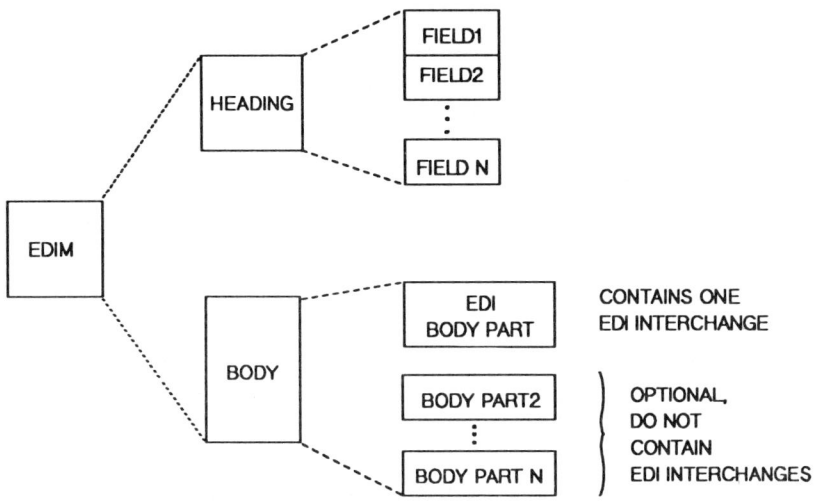

Figure 3.2 - Structure of EDIM

The key point to retain is that an EDIM consists of a *heading* and a *body*. The *heading* contains both X.400-specific data, and EDI Interchange-specific data. The *body* contains the EDI Interchange.

An EDIM contains at most one EDI Interchange. Alternative designs that allowed for more than one EDI Interchange within an EDIM were examined, but not adopted because of the increased complexity that would have been required in the protocol (for example, multiple *headings*).

3.2   The EDIM Heading

The EDIM Heading contains the information that the EDI UA needs in order to provide the services that can be requested by the EDI User. Annex B of F.435 defines in detail the features, functions and capabilities of P_{edi}; these definitions are given in terms of the X.400 concept of *Element of Service*. Most *Elements of Service* correspond to a *field* in the EDIM Heading. Annex L of X.435 shows the correspondence between *Elements of Service* and protocol elements in P_{edi}.

In this book, we shall focus on explaining the meaning and use of the EDIM Heading *fields*, with the understanding that, by doing this, we will at the same time explain the meaning and use of the *Elements of Service* defined in F.435.

The EDIM Heading is defined in clause 8.2 of X.435. It contains many *fields*. These can be grouped into two categories:

1.   X.400 *fields* such as *originator*

2.   EDI Interchange *fields* such as *interchange-sender*.

Some of the EDIM Heading Fields are closely related to the IPM Heading Fields, and do not require any special discussion. The descriptions given in clause 8.2 of X.435 are clear. The following are worth noting:

EDIM Identifier: is defined in clause 7.1 of X.435. It provides a unique identifier for each EDIM. That is, no EDIM has the same EDIM Identifier as any other EDIM.

Recipients: multiple *recipients* are allowed. Several *data elements* from the EDI *header segment* are copied into the EDIM Recipient Field. See section 3.3.

If the same ORName is specified for several *recipients*, and *notifications* are requested, then the *originator* may receive indistinguishable EDINs: see section 6.3.

NOTE: The *recipient field* contains an ORName. The definition of ORName is *imported* from X.411, which states that the ORName can consist of either a Directory Name or an ORAddress or both. If the ORName consists of a Directory Name only then the ORAddress is supplied by the MTA.

EDIN Receiver: is crucial to *forwarding* and *EDI Notifications* (EDINs). EDINs will be discussed in Chapter 4. The use of the EDIN Receiver *field* will be discussed in more detail in Chapters 5 and 6.

The UA that creates an EDIM can place an arbitrary ORName in the EDIN Receiver Field. This allows an organization to designate a specific UA to act in a specialist function to process all incoming EDINs. In other words, a UA is not obliged to receive itself the EDINs that it has requested.

When a UA *forwards* an EDIM, it must follow certain specific rules that specify the value of the EDIN Receiver Field. These rules are explained in Chapters 5 and 6.

EDI Body Part Type: an *object identifier* that indicates which EDI syntax is used for the EDI Interchange (for example, EDIFACT, ANSI X12, UN/TDI, private), and which character set is used to encode the EDI Interchange (ASCII 7-bit, ASCII 8-bit, EBCDIC, arbitrary binary).

EDI Message Type: a set of *octet strings* that indicates which types of *EDI Messages* (ANSI *Transaction Sets)* are contained in the EDI Interchange (for example, orders, invoices).

Responsibility Forwarded: this *field* can have one of the two values "true" or "false". Absence of the *field* is equivalent to the value "false". The use of this field is discussed in Chapter 5.

3.3   Relation of EDIM Heading to EDI Interchange

Several of the EDIM Heading Fields are copied from *data elements* contained in the EDI Interchange *header segment*, for example, the EDIFACT UNB *segment*.

For convenience, X.435 consistently refers to *data elements* defined in the EDIFACT UNB *segment*. No loss of generality is implied, since other EDI standards have comparable *header segments* and *data elements*. Annex K of X.435 shows how *data elements* from other syntaxes should be copied into the EDIM Heading.

The reason for copying data from the EDI Interchange Header Segment to the EDIM Heading is to allow the EDI UA to make decisions on the basis of that data, without having to parse the EDI Interchange itself. The EDI UA must be able to parse ASN.1; by copying data from the EDI Interchange Header Segment into the EDI Heading we avoid requiring that the EDI UA also be able to parse EDI syntaxes (for example, EDIFACT, ANSI X12).

A requirement to be able to parse EDI syntaxes would significantly complicate the implementation of an EDI UA that can make routing and forwarding decisions on the basis of data contained in the EDI Interchange Header Segment.

The following are worth noting:

Recipients: several EDI Header Segment Data Elements can be specified on a per-*recipient* basis. These *fields* are defined in clause 8.2.3 of X.435.

The EDI Interchange is identical for all *recipients*. Therefore, the data contained in the EDIM Recipients *field* override the data contained in the EDI Interchange Header Segment, since, for a given *recipient*, the data contained in the EDIM Recipients *field* may not be identical to the data contained in the EDI Interchange Header Segment.

This feature has been included in the P_{edi} protocol in order to support future possible EDI syntax standards which will either allow multiple recipients to be specified in the EDI Interchange Header Segment, or will not require an EDI Interchange Header Segment if X.400 is used as the data transmission protocol.

See also section 15.2.

Action Request: this replaces the IPM concepts of Primary and Copy Recipients. The P_{edi} protocol specifies specific actions that are requested from *recipients*. See section 12.1.

Notification Requests: in the P_{edi} protocol *notifications* are very different from IPM Notifications.

EDIM Notifications will be discussed in Chapters 4 and 6.

EDIM Security Requests and Notifications will be discussed in chapter 10.

Example 3.1 shows an EDIM and the EDI Interchange contained in it.

Example 3.1 - EDIM

EDIM Heading	
Field	Value
This EDIM	CH/ARCOM/HP/EDI/PURCH, 90/01/26/10:21:04
Originator	CH/ARCOM/HP/EDI/PURCH

Recipients
  Recipient
  Notification Requests
  Interchange Recipient
  Interchange Control Reference
	
	CH/ARCOM/SUPPLIER
	NN
	SUPPLIER
	135

EDI Body Part Type	EDIFACT, ISO 646
EDI Message Type	ORDERS
Interchange Sender	HP
EDIM Body	
UNB+UNOA:1+HP+SUPPLIER+90/01/26+135' UNH+1+ORDERS:1' ... UNZ+1+135'	

3.3.1   Proprietary EDI syntax

The P_{edi} protocol allows proprietary EDI syntaxes to be used. That is, the EDI Interchange contained in an EDIM may be structured according to a proprietary syntax convention.

In this case, the EDI user has the responsibility of giving enough information to the EDI UA so that the UA can create the appropriate EDIM Heading Fields. For example, a proprietary syntax may have an equivalent of the EDIFACT UNB Sender *data element*, and, if so, the EDI UA should place this equivalent in the Interchange Sender Field of the EDIM Heading.

3.4   Body of EDIM forwarded unchanged

An EDIM may be *forwarded* with no changes. In this case, the EDIM, and optionally its P1 *envelope*, is simply packed into a *body part*, which becomes the only *body part* of a new EDIM that is *submitted*. Figure 3.3 illustrates this.

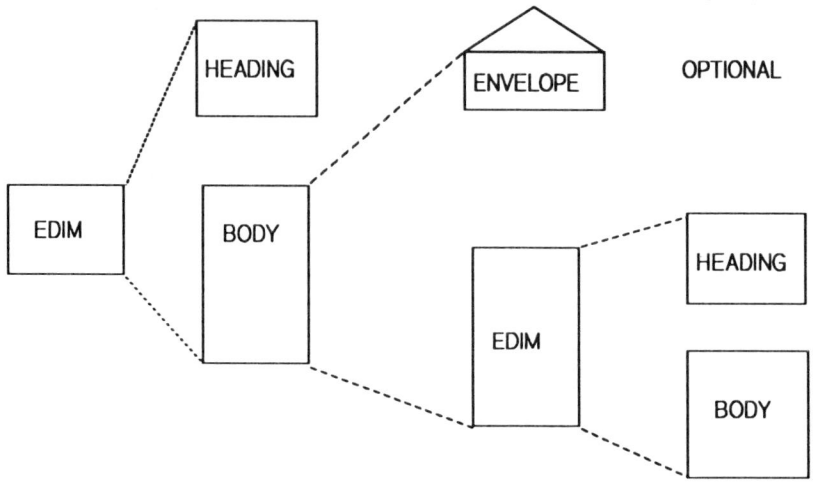

Figure 3.3 - Forwarding unchanged

The *forwarding* action imposes certain rules on the UA with respect to the EDIM Heading Fields of the *forwarded* EDIM. These rules will be discussed in Chapter 5.

In addition, clause 17.3.3.1 of X.435 specifies that the P1 *envelope* <u>must</u> be included in the *forwarded* EDIM if security features are requested (see Chapter 10) or if the EDIM to be forwarded does not itself contain a *forwarded* EDIM. That is, the P1 *envelope* must be included the first time forwarding takes place.

The P1 *envelope* is required for certain security services, so it is clear that it must be *forwarded* if security is required.

In addition, mandatory inclusion of the P1 *envelope* the first time an EDIM is forwarded facilitates tracking and logging, since the P1 *envelope* contains addressing information not found in the EDIM itself.

3.5 Body of EDIM forwarded with changes

No changes may be made to the *body parts* of an EDIM. However, *body parts* may be added to a *forwarded* EDIM, and the *forwarded* EDIM might not contain all of the *body parts* of the original EDIM.

If *body parts* are added to a *forwarded* EDIM, they appear as additional *body parts* after the *body part* that contains the *forwarded* EDIM.

If *body parts* are dropped from the *forwarded* EDIM, they are replaced by a place holder. The place holder mechanism is defined in clause 8.3.2 of X.435.

A *body part* that contains a *forwarded* EDIM may not be dropped. One reason for this restriction is that the original EDIM Heading is contained within the *forwarded EDIM body part*, and, if it were dropped, the Cross Referencing Information (see Chapter 8) would be lost. Another reason is that the *forwarded EDIM body part* might contain the P1 *envelope* needed for security services (see Chapter 10).

If it is necessary to drop a *body part* that contains a *forwarded* EDIM, then the UA must pass the EDIM out of the X.400 world, create a new *message* containing the *body parts* to be sent, and *submit* this new *message*. In this case, the new *message* is, in X.400 terms, completely unrelated to the EDIM received.

Rules related to dropping *body parts* are given in clauses 8.3.2 and 17.3.3.2 of X.435.

Figure 3.4 illustrates the case of an EDIM that is *forwarded* with both added and dropped (*removed*) body parts. A more detailed illustration is given in Figure 2 of clause 8.3.2 of X.435.

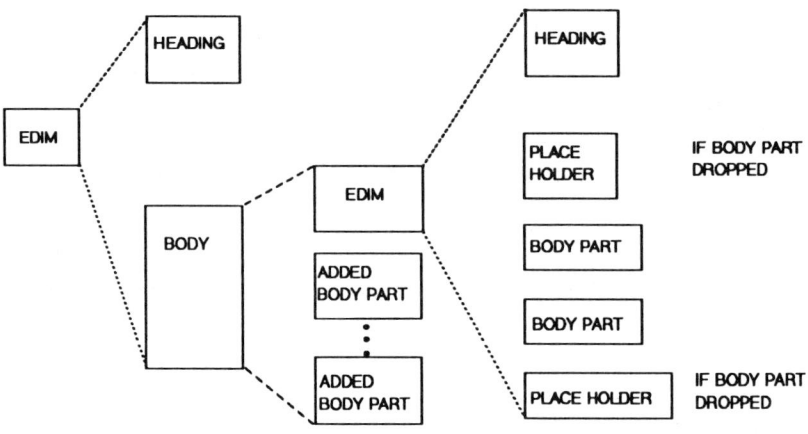

Figure 3.4 - Forwarding with changes

The *forwarding* action imposes certain rules on the UA with respect to
the EDIM Heading Fields of the *forwarded* EDIM. These rules will be
discussed in Chapter 5.

4   Structure of an EDI Notification

If the EDI Message is the heart of the P_{edi} protocol, then the *EDI Notification* (EDIN) is the brain of the P_{edi} protocol. The EDI Notification is very different from the IPM Notification.

The general concept of an EDI Notification is simple: it is an acknowledgement that the *recipient* UA sends to the *originator* UA of an EDIM. The *recipient* UA sends the EDIN only if it is requested to do so by the *originator* UA. The exact circumstances under which a *recipient* UA sends an EDI Notification, and what type of EDI Notification it sends, will be discussed in Chapter 6.

EDINs cannot be *forwarded*, and EDI Notifications cannot be requested for EDINs. There are three types of EDINs:

NN: Negative Notification, used when the *recipient* UA rejects responsibility for the received EDIM.

PN: Positive Notification, used when the *recipient* UA accepts responsibility for the received EDIM.

FN: Forwarding Notification, used when the *recipient* UA does not accept responsibility for the received EDIM, and *forwards* both responsibility and the EDIM.

Clause 9 of X.435 defines the EDIN, and Figure 3 of the clause illustrates the structure of an EDIN. Figure 4.1 illustrates the key features of an EDIN.

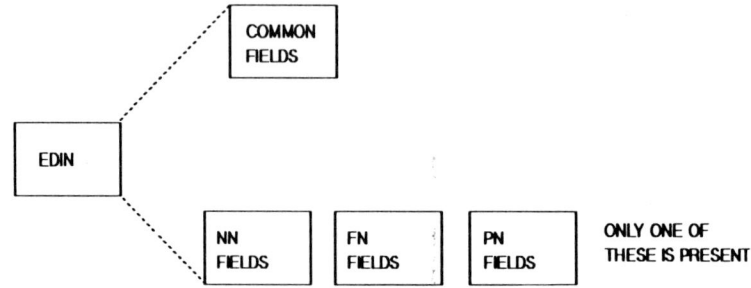

Figure 4.1 - Structure of EDIN

All EDINs contain certain *common fields*. In addition, each type of EDIN contains *fields* particular to that type of EDIN.

The EDIN Common Fields are defined in clause 9.1 of X.435. The following are worth noting:

Subject EDIM: contains the EDIM Identifier of the EDIM which has been received, and for which the EDIN (the notification) is being sent.

EDIN Originator: contains the ORName of the UA which sends the EDIN.

First Recipient: contains the ORName of the UA to which the EDIM was first sent by the original *originator*. This *field* is of crucial importance when processing received EDINs, as we will see in Chapter 6.

Security Elements: will be discussed in Chapter 10.

EDIN Initiator: will be discussed in section 12.2.

Detailed rules for setting the Subject EDIM and First Recipient *fields* of the EDIN using the sub-*fields* contained in the EDIN Receiver *field* of the EDIM are given in clause 17.3.1.1 of X.435. These rules are explained in section 6.2.

The *fields* that are specific to each type of EDIN consist essentially of the reason code (a code that indicates the reason for the action taken by the recipient UA). Clauses 9.3.1 and 9.4.2 of X.435 define the NN Reason Codes and the FN Reason Codes.

NN Reason Codes fall into one of three categories:

- codes generated by the UA

- codes generated by the EDI application (referred to as *user* in the standards)

- codes generated by a physical delivery access unit (see section 15.6)

Codes generated by the UA fall into one of two categories:

- basic codes, as specified in Annex B of F.435 for the Element of Service "EDI Notification Request"

- diagnostic codes, which include additional information

Codes generated by the EDI application (*user*) fall into one of two categories:

- basic codes, as specified in 9.3.1 of X.435

- diagnostic codes, which are not specified in X.435 (the EDI application is free to specify any integer value in this *field*).

See clause 9.3.1 of X.435 for detailed lists of allowed code values.

The structure of FN Reason Codes is essentially identical to that for NN Reason Codes, and detailed lists of allowed code values are contained in clause 9.4.2 of X.435. (See sub-section 10.5.3 for use of the FNUAMS Security Check *field*).

For all types of notfications, including PN Notifications, additional information can be given in the EDI Supplementary Information *field*, which can be an arbitrary string of text (IA5String).

Example 4.1 shows what an EDIN would look like for the EDIM shown as Example 3.1 of section 3.3, if the EDIM could not be accepted for an unknown reason.

Example 4.1 - EDIN

EDIN Common Fields	
Field	Value
Subject EDIM	CH/ARCOM/HP/EDI/PURCH, 90/01/26/10:21:04
EDIN Originator	CH/ARCOM/SUPPLIER
First Recipient	CH/ARCOM/SUPPLIER
Notification Time	90/01/26/12:43:54
EDIN Negative Notification Fields	
Field	Value
NN Reason Code NNUAMS Basic Code	UNSPECIFIED

4.1   Billing for EDINs

It would appear logical and desirable for the requestor of an EDIN to pay for the transmission of the EDIN.

Due to limitations in the P1 protocol, the P_{edi} protocol does not make any statements with respect to who should pay for transmission of EDINs. The entire question of billing for EDINs is matter for further study.

5   Forwarding

If the EDIM is the heart of the P_{edi} protocol and the EDIN its brain, then *forwarding* is its soul. *EDI Forwarding* is related to IPM Forwarding, but differs significantly because of the differences between EDI Notifications and IPM Notifications.

In order to understand EDI Forwarding, we will describe three different forwarding scenarios, referred to as Case 1, Case 2 and Case 3.

5.1   Case 1: no forwarding

In this scenario there is no forwarding. Figure 5.1 illustrates the scenario.

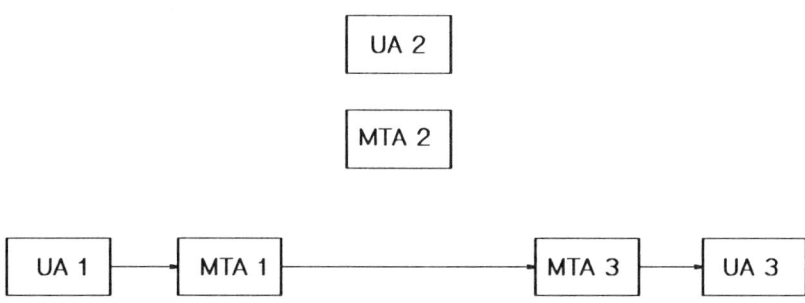

Figure 5.1 - No forwarding

The EDIM is *submitted* by UA1 to MTA1, *relayed* to MTA3 and *delivered* to UA3.

5.2   Case 2: responsibility not accepted

In this scenario both the EDIM, and responsibility for the EDIM, are forwarded. These actions can be performed only if the Responsibility Passing Allowed *field* of the EDIM Heading of the received EDIM holds the value "true", which is <u>not</u> the default value (see clause 8.2.3.4 of X.435). Figure 5.2 illustrates the scenario.

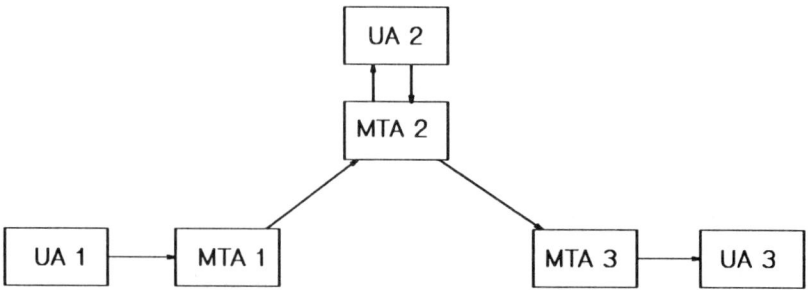

Figure 5.2 - Forwarding, responsibility not accepted

The EDIM is *submitted* by UA1 to MTA1, *relayed* to MTA2, *delivered* to UA2, *forwarded* to UA3. When UA2 *forwards* the EDIM, it *submits* the *forwarded* EDIM to MTA2, who *relays* it to MTA3, who *delivers* it to UA3.

UA2 must not modify in any way the EDIM that is *delivered* to it and *submitted* by it. That is, the EDIM that has been received is packed unchanged into the first, and only, *body part* of the EDIM that UA2 *submits* to MTA2.

If UA1 has requested *notification* from UA2, then UA2 must request *notification* from UA3, and must place the ORName of UA1 (or, more precisely, the ORName that appears in the EDIN Receiver Field of the EDIM received from UA1) in the *EDIN Receiver Field* of the *forwarded* EDIM. As we will see in Chapter 6, these rules guarantee that UA1 will receive the EDINs that it has requested.

The rules are given in detail in clauses 17.3.3.1 and 17.3.3.4 of X.435.

The *forwarded* EDIM must carry the value "true" in the Responsibility Forwarded *field* of its EDIM Heading.

5.3   Case 3: responsibility accepted

In this scenario the EDIM is forwarded, but responsibility is <u>not</u> forwarded. That is, the recipient UA accepts responsibility for the EDIM, but *forwards* the EDIM. Figures 5.3 and 5.4 illustrate the scenario.

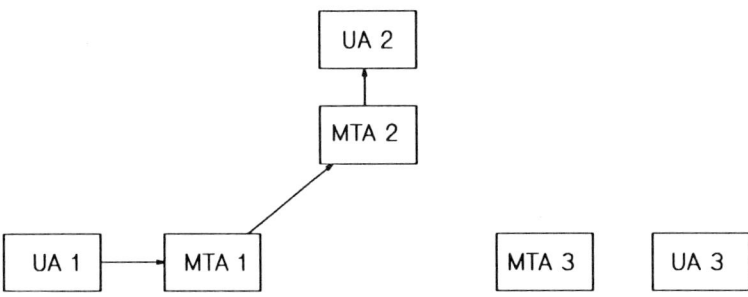

Figure 5.3 - Forwarding, responsibility accepted

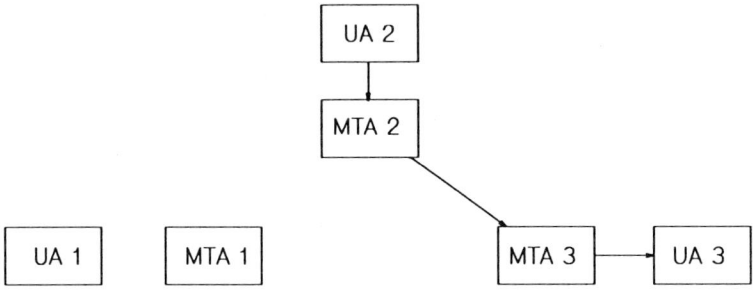

Figure 5.4 - Forwarding, responsibility accepted

In Figure 5.3 the EDIM is *submitted* by UA1 to MTA1, *relayed* to MTA2 and *delivered* to UA2. UA2 accepts responsibility (the mechanism by which UA2 accepts responsibility will be described in Chapter 6), and then *forwards* the EDIM to UA3.

Figure 5.4 illustrates the EDI Forwarding. UA2 *submits* the EDIM to MTA2, who *relays* it to MTA3, who *delivers* it to UA3.

UA2 must not modify in any way the *body parts* of the EDIM that is *delivered* to it and *submitted* by it. UA2 may add or drop *body parts*. That is, some, or all, of *body parts* of the EDIM that has been received are packed unchanged into the first *body part* of the EDIM that UA2 *submits* to MTA2. Additional *body parts* may be added to the EDIM that UA2 *submits* to MTA2.

If UA1 has requested *notification* from UA2, then UA2 must send an EDIN to UA1. UA2 has the option to request *notification* from UA3; however, if it does so, it must not place the ORName of UA1 in the EDIN Receiver Fields of the *forwarded* EDIM. As we will see in Chapter 6, these rules guarantee that UA1 will not receive EDINs that it has not requested.

The rules are given in detail in clauses 17.3.3.2 and 17.3.3.4 of X.435.

The Responsibility Forwarded *field* is not present in the EDIM Heading of the *forwarded* EDIM; its absence is equivalent to the value "false".

5.4   Summary: rules when forwarding

5.4.1   Heading

To summarize, the following rules apply to the EDI Heading Fields of a *forwarded* EDIM:

EDIMIdentifier: new value created by the UA that performs the *EDI forwarding* operation.

Originator: ORName of the UA that performs the *EDI forwarding* operation.

Recipients: created by the UA that performs the *EDI forwarding* operation. If the EDIM that is being *forwarded* had a *notification request*, and responsibility is forwarded, then one, and only one, of the new *recipients* must contain the identical *notification request*.

EDIN Receiver: if responsibility is forwarded, the EDIN Receiver of the *forwarded* EDIM must be set to the ORName contained in the EDIN Receiver of the EDIM that is being *forwarded*.

                          If responsibility is <u>not</u> forwarded, the EDIN Receiver of the *forwarded* EDIM can be set to any value, <u>except</u> the ORName contained in the EDIN Receiver of the EDIM that is being *forwarded*.

                          Detailed rules concerning how to set the sub-*fields* of the EDIN Receiver *field* are contained in clause 17.3.3.4 of X.435. These rules are explained in detail in sub-section 5.4.2.

Responsibility Forwarded: "true" if responsibility is forwarded, not present otherwise.

Other Heading Fields: if responsibility is forwarded, only those EDIM Heading *fields* that differ from their default values are copied to the new EDIM Heading (see the general rules in clause 17.3.3 of X.435 and the specific rules in clause 17.3.3.4).

If responsibility is <u>not</u> forwarded, there are no restrictions concerning EDIM Heading Fields of the new EDIM.

Table 5.1 summarizes the setting of the most important EDIM Heading Fields.

Table 5.1 - EDIM Heading Fields when forwarding

	Responsibility forwarded	Responsibility <u>not</u> forwarded
EDIM Identifier	new value	new value
Notification request	unchanged for one recipient	any value
EDIN Receiver	unchanged	not same value as in original EDIM
Responsibility Forwarded	TRUE	not present

NOTE: when *responsibility* is <u>not</u> forwarded, X.435 does not actually prohibit setting the EDIN Receiver to the same value contained in the original EDIM.

However, it should be clear that, in most cases, the EDIN Receiver should be set to the ORName of the UA that accepts responsibility for the original EDIM, and thus should <u>not</u> be set equal to the value contained in the original EDIM.

See Figure 6.2 in Chapter 6.

5.4.2   EDIN Receiver

Clause 17.3.3.4 of X.435 specifies rules for creating the sub-*fields* of the EDIN Receiver *field* of the EDIM when *forwarding*.

The purpose of these rules, and the purpose of the sub-*fields* of the EDIN Receiver *field*, is to make it easy for a *recipient* UA to generate an EDIN if the *recipient* does not *forward* the EDIM. A *forwarding* UA must perform a certain amount of extra work, in order to facilitate the *recipient* UA's task of generating EDINs.

The rules relate to the EDIN Receiver *field* of the new EDIM Heading created by the *forwarding* UA. The EDIN Receiver *field* contained in the old EDIM Heading which is packed into the *forwarded body part* is not changed in any way.

Forwarding when responsibility is accepted

Creation of the EDIN Receiver *field* is completely optional, just as it is when first creating an EDIM. Normally, there is no reason to create this field. If EDINs are requested, and if the *originator* UA does not wish to receive the EDINs, then the EDIN Receiver sub-*field* can be set to the ORName of the UA which should receive and process the EDINs. The rules are:

EDIN Receiver: ORName of the UA which should receive the requested EDINs, if this UA is not the *originator*.

Original EDIM Identifier: not present.

First Recipient: not present.

Forwarding when responsibility is not accepted

Creation of the EDIN Receiver *field* is mandatory, and the *field* <u>must</u> contain all the optional sub-*fields*. This makes it easy for the *recipient* to generate the EDINs, without having to unpack the nested structure of the *forwarded* EDIM.

Since *notification* requests (if any are present) are *forwarded* to one and only one *recipient*, the forwarded EDIM contains a *notification* request for at most one *recipient*, and the First Recipient sub-*field* can be present, in accordance with the rules contained in 8.2.4 of X.435.

If the EDIM being *forwarded* already contains an EDIN Receiver *field* then the *forwarding* UA must use the values contained therein when creating the EDIN Receiver *field* of the *forwarded* EDIM.

If the EDIM being *forwarded* does not contain all the sub-*fields* of the EDIN Receiver *field*, then the following rules must be used:

EDIN Receiver: ORName of the *originator* of the EDIM being *forwarded* (or of some other UA that should receive the EDINs).

Original EDIM Identifier: EDIM Identifier of the EDIM being *forwarded*.

First Recipient: ORName of the UA which the *originator* specified as *recipient*. This ORName is normally the ORName of the UA which received the EDIM. An exception occurs if an MTA has performed *redirection*. In case of *redirection*, the *recipient* UA must use the P1 *envelope* to create the correct First Recipient ORName.

Summary

So long as no UA forwards responsibility, the EDIN Receiver *field* will normally not be present, or will contain only the ORName of the UA to which EDINs should be sent, if the *originator* UA does not want to receive the EDINs.

The first UA that forwards responsibility has the obligation to create all the sub-*fields* of the EDIN Receiver *field*, in such a way that the final *recipient* will be able to use this *field* to generate the EDIN. Subsequent UA's that forward responsibility simply copy the EDIN Receiver *field* unchanged.

"An exception occurs if an MTA has performed redirection or DL-expansion. In these cases, the recipient UA must use the P1 envelope to create the First Recipient ORName (see clause D.3.1.1 of X435)

Examples 5.2 and 5.3 illustrate the rules. In both examples UA1 creates an EDIM and sends it to UA2, who forwards it to UA3. Example 5.1 illustrates the EDIM Heading created by UA1. These values are used in Examples 5.2 and 5.3. In Example 5.2 UA2 accepts responsibility. In Example 5.3 UA2 forwards responsibility.

Example 5.1 - Original EDIM Heading

EDIM Heading created by UA1	
Field	Value
This EDIM	CH/ARCOM/HP/EDI/PURCH, 90/01/26/10:21:04
Originator	CH/ARCOM/HP/EDI/PURCH

Recipients
  Recipient
  Notification Requests
	CH/ARCOM/SUPPLIER
	PN

EDIN Receiver EDIN Receiver	CH/ARCOM/HP/EDI/EDIN_R

Example 5.2 - Forwarded EDIM Heading, Responsibility Accepted

EDIM Heading created by UA2	
Field	Value
This EDIM	CH/ARCOM/SUPPLIER, 90/01/26/11:18:53
Originator	CH/ARCOM/SUPPLIER

Recipients
  Recipient
  Notification Requests
	CH/ARCOM/SUB_CONTRACT
	PN

EDIN Receiver EDIN Receiver	CH/ARCOM/SUPPLIER/EDIN

Example 5.3 - Forwarded EDIM Heading, Responsibility Forwarded

EDIM Heading created by UA2	
Field	Value
This EDIM	CH/ARCOM/SUPPLIER, 90/01/26/11:18:53
Originator	CH/ARCOM/SUPPLIER

Recipients
  Recipient
  Notification Requests
	
	CH/ARCOM/SUB_CONTRACT
	PN

EDIN Receiver
  EDIN Receiver
  Original EDIM Identifier
 	
  First Recipient
	
	CH/ARCOM/HP/EDI/EDIN_R
	CH/ARCOM/HP/EDI/PURCH,
	90/01/26/10:21:04
	CH/ARCOM/SUPPLIER

Responsibility Forwarded	TRUE

5.4.3   Body

The following rules apply to the *body* of a *forwarded* EDIM:

Responsibility forwarded: the original EDIM is packed unchanged in the first, and only, *body part* of the *forwarded* EDIM.

Responsibility not forwarded: some, or all, of the *body parts* of the original EDIM are packed unchanged in the first *body part* of the *forwarded* EDIM.   Other *body parts* may be added.

5.4.4   UA operation

Taken together, various clauses of X.435 restrict and specify the operation of a UA with respect to *forwarding* and sending EDI Notifications.   See Chapter 6 for a more complete discussion of sending EDI Notifications.

When reading the standard, it is not immediately clear how the different key sentences apply, nor is the flow of logic implied by the language immediately obvious.   The annotated flow charts which follow (Figures 5.5

and 5.6) summarize the operations of a UA, and point to the relevant clauses of X.435.

The flow chart in Figure 5.5 shows the basic operations for a UA. The numbers which appear next to the boxes in the figure refer to clauses in X.435.

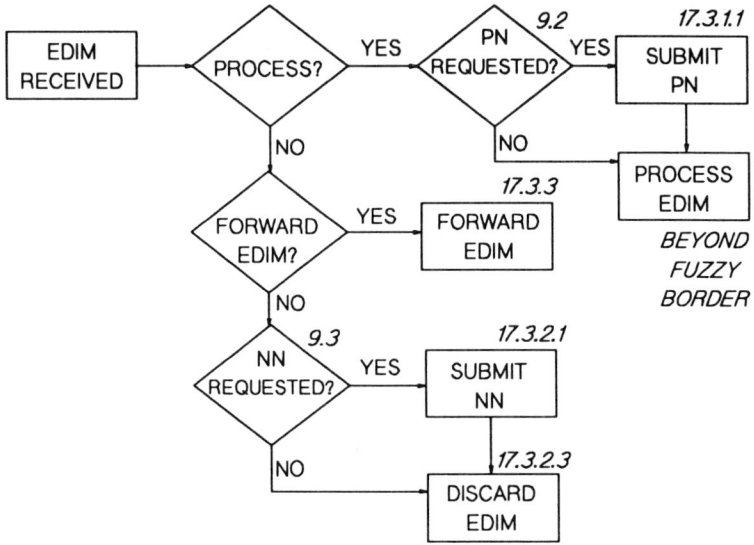

Figure 5.5 - UA operation: flow chart 1, basic

NOTE - The notation "beyond fuzzy border" in Figure 5.5 means that the standards do not specify how an EDIM should be processed. Such matters are beyond the scope of the standards. See section 15.4 for a more complete discussion of this point.

Forwarding is a completely optional feature of a UA, and a UA is not obliged to implement *forwarding*. In this case, the box marked "forward EDIM?" would always be left through the "no" branch.

If a UA implements *forwarding*, its operation is further restricted by several clauses in X.435. The flow chart in Figure 5.6 shows the operations for a UA that implements *forwarding*. The numbers which appear next to the boxes in the figure refer to clauses in X.435. See Chapter 6 for a discussion of the concept of *responsibility* and for a complete discussion of sending EDI Notifications.

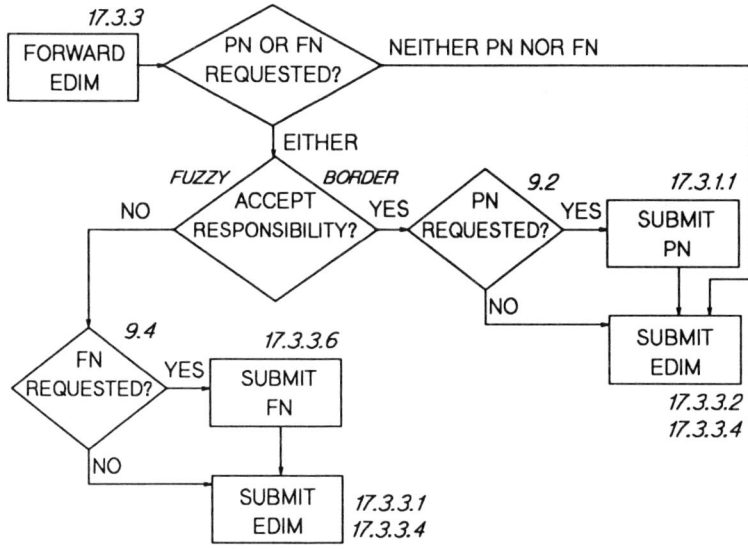

Figure 5.6 - UA operation: flow chart 2, forwarding

NOTE - The notation "fuzzy border" in Figure 5.6 means
that the standards do not specify exactly when or why
responsibility should be accepted or not accepted. Such
matters are beyond the scope of the standards. See section
15.4 for a more complete discussion of this point.

One key point illustrated in Figure 5.6 is that the rules governing
construction and submission of an EDIM when *forwarding* are different
according to whether *responsibility* has been accepted or not. Clause 17.3.3.1
and the first part of clause 17.3.3.4 of X.435 cover the rules if *responsibility* is
not accepted. Clause 17.3.3.2 and the second part of clause 17.3.3.4 of X.435
cover the rules if *responsibility* is accepted. The application of these rules is
discussed in sub-sections 5.4.1, 5.4.2 and 5.4.3.

The following points are worth noting:

- If only FN is requested, and the *forwarding* UA accepts
  responsibility, then the UA that requested the FN will not
  receive any EDI Notifications.

- If only PN is requested, and the *forwarding* UA does <u>not</u> accept *responsibility*, then the *forwarding* UA forwards the PN request together with the *forwarded* EDIM, as discussed in sub-section 5.4.1.

5.5   Loop prevention

It is clear that a *forwarding* loop might occur. For example, UA1 *forwards* to UA2, who *forwards* to UA3, who *forwards* to UA1.

In order to prevent infinite loops, the P_{edi} protocol specifies that a UA must not *forward* an EDIM that contains a *forwarded* EDIM that contains an EDIM Identifier that contains the ORName of the UA itself.

That is, if a UA detects that it has already *forwarded* an EDIM, it will not *forward* it again, and this prevents infinite looping.

Clause 17.3.3.3 of X.435 defines in detail the loop prevention mechanism.

6 Responsibility: sending and receiving notifications

The P_{edi} protocol includes a concept called *EDI Responsibility*. This concept is defined in clause 8.1 of F.435.

EDI Responsibility is closely tied to EDINs. The concept of an *EDI Notification* (as opposed to the structure of an EDIN) is defined in clause 7.5 of F.435.

6.1 Definition of EDI Responsibility

An alternative definition of *EDI Responsibility* can be given purely in terms of sending and receiving EDI Notifications. Combining clauses 7.5 and 8.1 of F.435, the alternative definition is:

6.1.1 Clauses 7.5 and 8.1 of F.435 are equivalent to the following:

The EDIMS includes objects called EDI Notifications, and it includes the concept of forwarding.

Forwarding is performed by a UA or MS. Body parts that are forwarded cannot be changed in any way. Body parts may be added to or dropped from a forwarded EDIM. If body parts are added or dropped and if the original originator requested positive EDI notifications, then a Positive Notification must be generated.

EDI Notifications are used to allow end-to-end acknowledgements of receipt when messages are forwarded. EDI Notifications take three forms:

- Forwarding Notification (FN)

- Negative Notification (NN)

- Positive Notification (PN)

When requested to do so by the originator, a receiving UA generates notifications as follows:

- Forwarding Notification whenever the receiving UA forwards the EDIM, and no previous UA has sent a Positive or Negative Notification, and no body parts are added or dropped. The Forwarding Notification contains the name of the UA to which the EDIM has been forwarded.

- Negative Notification whenever the receiving UA rejects the EDIM. The reason for rejection is contained in the Negative Notification.

- Positive Notification whenever the receiving UA does any of the following:

> adds or drops body parts from the EDIM when forwarding;

> passes the EDIM to the EDI user;

> decides to forward and wishes to prevent further EDINs from being sent to the originator. In this case, the UA that sends the Positive Notification is at the end of the forwarding chain that started with the original originator, and the original originator is made aware of this fact when it receives the Positive Notification.

The originator has the option to request any, or any combination, of FN, NN and PN. No notifications shall be sent unless requested.

6.1.2   Summary definition

To put it in a nutshell, if we assume that PN and NN Notifications are requested, we can say that a UA accepts *EDI Responsibility* if and only if it sends a PN EDIN, refuses *EDI Responsibility* if and only if it sends an NN EDIN, and forwards *EDI Responsibility* if and only if it *forwards* an EDIM without sending either an NN or PN EDIN.

6.2   Sending notifications

We are now in a position to understand why, how and when EDI Notifications are sent by a recipient UA, and to whom they are sent. See also section 5.4.4 for an overall view of UA operations, and descriptive flow charts that show when EDI Notifications are sent.

A UA sends an FN EDIN only if the FN value is set in the Notification Request Field of the received EDIM, and the UA does not accept responsibility for the EDIM, and the UA *forwards* the EDIM.

A UA sends an NN EDIN only if the NN value is set in the Notification Request Field of the received EDIM, and the UA refuses to accept responsibility for the EDIM.

A UA sends an PN EDIN only if the PN value is set in the Notification Request Field of the received EDIM, and the UA accepts responsibility for the EDIM.

Table 6.1 summarizes the situation:

Table 6.1 - Sending EDINs

	Responsibility forwarded	Responsibility not forwarded
FN requested	FN sent	no FN sent
NN requested	no NN sent	NN sent if responsibility is refused
PN requested	no PN sent	PN sent if responsibility is accepted

NOTE: If responsibility is forwarded, and an FN EDIN was requested, the receiving UA must send the FN at once. If it subsequently receives a *non-delivery report* from the MTA to whom it *submitted* the *forwarded* EDIM, the UA has the option to send an NN in addition to the FN. This action is described in bullet c of clause 17.3.3.1 of X.435.

The reason why sending the NN is optional in case of *non-delivery*, rather than mandatory, is that if the NN were mandatory, then any UA that *forwards* an EDIM would have to keep track of all *forwarded* EDIMs, in order to be able to create the NN EDIN if a *non-delivery* report is received from the MTA. The *non-delivery* report does not contain enough information to create the EDIN.

In all cases, the EDINs are sent to the ORName contained in the EDIN Receiver Field of the received EDIM, if the EDIN Receiver Field is present. If the EDIN Receiver Field is not present, the EDINs are sent to the ORName contained in the Originator Field of the received EDIM.

By referring to Table 5.1 of Chapter 5, we can see that, if the received EDIM has already been *forwarded*, the EDIN Receiver Field contains the ORName of the EDIM in the *forwarding* chain that had accepted responsibility, so that the EDIN sent to the ORName specified in the EDIN Receiver is indeed sent to the UA that is expecting an EDIN.

Figure 6.1 illustrates the situation if responsibility has been forwarded (as discussed in section 5.2). These actions can be performed only if the Responsibility Passing Allowed *field* of the EDIM Heading of the received EDIM holds the value "true", which is <u>not</u> the default value (see clause 8.2.3.4 of X.435).

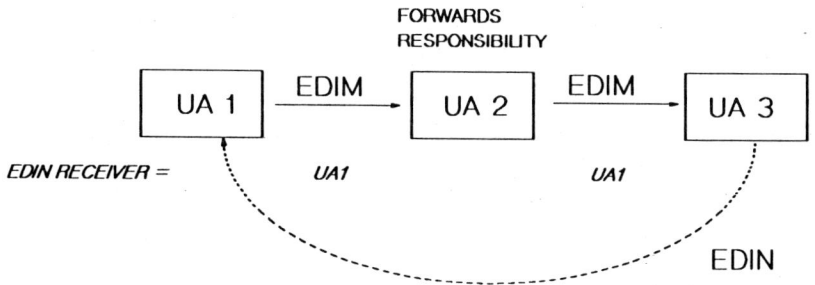

Figure 6.1 - Responsibility forwarded

UA1 creates an EDIM, requests PN and NN *notifications* and sets the EDIN Receiver Field to the ORName of UA1. UA1 sends this EDIM to UA2. UA2 does not accept responsibility, and *forwards* the EDIM, and responsibility, to UA3. UA2 sets the EDIN Receiver Field of the *forwarded* EDIM to the ORName of UA1. Thus, when UA3 accepts (or refuses) responsibility, and sends the EDIN, UA3 will send the EDIN to UA1, as desired.

Figure 6.2 illustrates the situation if responsibility has not been forwarded (as discussed in section 5.3).

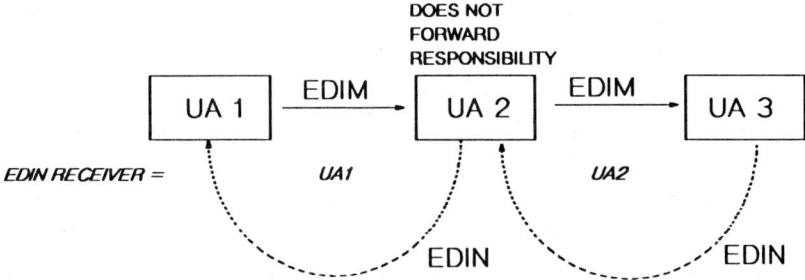

Figure 6.2 - Responsibility not forwarded

UA1 creates an EDIM, requests PN and NN *notifications* and sets the EDIN Receiver Field to the ORName of UA1. UA1 sends this EDIM to UA2. UA2 accepts responsibility, and sends an EDIN to UA1. UA2 also *forwards* the EDIM to UA3, requesting PN and NN *notifications*. UA2 sets the EDIN Receiver Field of the *forwarded* EDIM to the ORName of UA2. Thus, when UA3 accepts (or refuses) responsibility, and sends the EDIN, UA3 will send the EDIN to UA2, as desired.

Detailed rules for setting the Subject EDIM and First Recipient *fields* of the EDIN using the sub-*fields* contained in the EDIN Receiver *field* of the EDIM are given in clause 17.3.1.1 of X.435.

If the EDIN Receiver *field* is present in an EDIM, then the EDIN *fields* are set according to the following rules:

Subject EDIM: value contained in the Original EDIM Identifier sub-*field* of the EDIN Receiver *field* of the received EDIM.

First Recipient: value contained in the First Recipient sub-*field* of the EDIN Receiver *field* of the received EDIM.

If the EDIN Receiver *field* is <u>not</u> present in an EDIM, then the EDIN *fields* are set according to the following rules:

Subject EDIM: EDIM Identifier of the received EDIM.

First Recipient: ORName which the *originator* specified as *recipient*. This ORName is normally the ORName of the UA which received the EDIM. An exception occurs if an MTA has performed *redirection.* In cases of *redirection*, the *recipient* UA must use the P1 *envelope* to create the correct First Recipient ORName. (see clause 7.3.1.1 of X.435

The examples below illustrate the EDINs created by UA3 in response to the EDIMs of examples 5.2 and 5.3 of Chapter 5. In example 6.1, the EDIN is sent by UA3 to the EDIN Receiver specified by UA2, the UA that accepted responsibility. In example 6.2, the EDIN is sent by UA3 to the EDIN Receiver specified by UA1, the original *originator*.

Example 6.1 - EDIN, Responsibility Accepted

EDIN Common Fields	
Field	Value
Subject EDIM	CH/ARCOM/SUPPLIER, 90/01/26/11:18:53
EDIN Originator	CH/ARCOM/SUB_CONTRACT
First Recipient	CH/ARCOM/SUB_CONTRACT

Example 6.2 - EDIN, Responsibility Forwarded

EDIN Common Fields	
Field	Value
Subject EDIM	CH/ARCOM/HP/EDI/PURCH, 90/01/26/10:21:04
EDIN Originator	CH/ARCOM/SUB_CONTRACT
First Recipient	CH/ARCOM/SUPPLIER

At this point, the determined reader should turn to clauses 8.3, 8.4 and 8.5 of F.435, which provide a detailed illustration of the flow of EDIMs and EDINs in the three cases discussed in Chapter 5.

for example

NOTE: A UA may receive the same EDIM twice, due to re-transmission in case of suspected network problems.

As specified in clause 17.3 of X.435, a receiving UA may or may not be able to detect the reception of a duplicate EDIM. If it is able to detect the duplicate, and *notifications* are requested, then the UA shall send an NN EDIN for the duplicate.

If the UA is <u>not</u> able to detect the duplicate, and *notifications* are requested, then the UA will send a PN EDIN for the duplicate. In this case, the original *originator* may receive two PN EDINs for the same EDIM.

6.3 Receiving notifications

We are now in a position to understand why, how and when EDI Notifications are received, and by whom they have been sent.

A UA can receive an EDIN only if it requested *notifications*. *Notifications* are requested by setting the appropriate *field* in the *recipient field* of the EDIM Heading when the EDIM is created and *submitted*.

At most one NN or PN EDIN is received for each *recipient* for which an NN or PN *notification* was requested.

Note that the received NN or PN will not necessarily come from the *recipient* for which the *notification* was requested, since the EDIM might have been *forwarded*.

However, the ORName of the *recipient* for which *notification* was requested is contained in the *First Recipient Field* of the received EDIN.

NOTE: If the same ORName was specified for several *recipients*, and *notifications* were requested from each, then the *originator* will receive indistinguishable EDINs, since the EDINs received will have the same Subject EDIM and the same First Recipient.

If it is absolutely necessary to specify the same *recipient* ORAddress more than once for the same EDIM, then uniqueness can be achieved either by using different values in the Organizational Units *fields* of the ORName, or by using different Directory Names, which have the same ORAddress component (but specify different ORNames).

6.3.1   PN, NN

If both PN and NN *notification* was requested, then, normally, either an NN or a PN will be received. However, in exceptional cases, it is possible that no EDIN whatsoever will be received. This situation is best explained with an example: UA1 sends an EDIM to UA2, and requests PN and NN; UA2 forwards both the EDIM and responsibility to UA3, so that UA3 has the obligation to send the PN or NN to UA1; before UA3 can send the EDIN, it suffers a catastrophic, non-recoverable failure, so no EDIN is sent to UA1.

6.3.2   FN

If FN *notification* is requested and the EDIM is *forwarded* several times, several FNs will be received for each *recipient* for which an FN was requested.

   If the EDIM is forwarded, only one FN will come from the *recipient* for which the FN was requested. Additional FNs will come from *recipients* to whom the EDIM is forwarded. All received FNs will contain (in the First Recipient Field) the ORName of the first *recipient* for which the FN was requested.

6.3.3   FN and PN, NN

If FN and PN, or FN and NN, or FN and PN and NN *notification* is requested and the EDIM is *forwarded* several times, several FNs will be received for each *recipient* for which *notifications* were requested, and at most one NN or PN will be received for each *recipient* for which *notifications* were requested.

   Figure 6.3 illustrates the situation when a UA sends an EDIM to several *recipients*, and requests FN, PN and NN *notifications*.

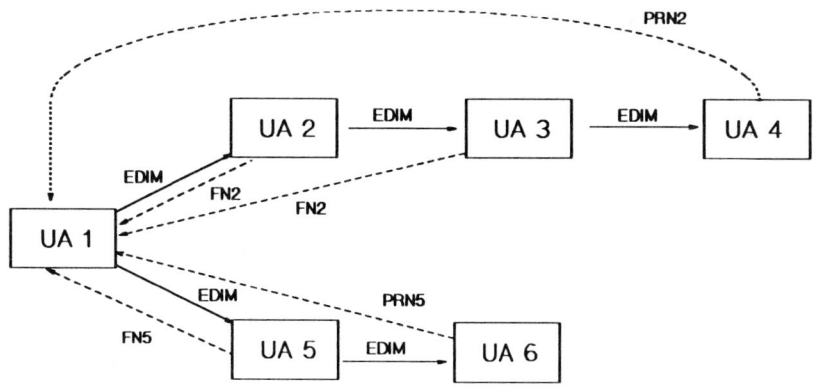

Figure 6.3 - Several notifications

UA1 sends an EDIM to <u>both</u> UA2 and UA5, and requests FN, PN and NN *notifications* from <u>both</u> UA2 and UA5.

UA2 forwards both the responsibility and the EDIM to UA3, and sends an FN to UA1. The FN contains the ORName of UA2 in the First Recipient Field (as indicated in Figure 6.3 by the fact that the FN is labelled FN2).

UA3 forwards both the responsibility and the EDIM to UA4, and sends an FN to UA1. The FN contains the ORName of UA2 in the First Recipient Field (as indicated Figure 6.3 by the fact that the FN is labelled FN2).

UA4 accepts responsibility, and sends a PN to UA1. The PN contains the ORName of UA2 in the First Recipient Field (as indicated Figure 6.3 by the fact that the PN is labelled PRN2).

UA5 forwards both the responsibility and the EDIM to UA6, and sends an FN to UA1. The FN contains the ORName of UA5 in the First Recipient Field (as indicated Figure 6.3 by the fact that the FN is labelled FN5).

UA6 accepts responsibility, and sends a PN to UA1. The PN contains the ORName of UA5 in the First Recipient Field (as indicated Figure 6.3 by the fact that the PN is labelled PRN5).

Thus, UA1 receives a total of five EDINs for one EDIM: three FNs and two PNs. Three EDINs contain the ORName of UA2 in the First Recipient Field, while two EDINs contain the ORName of UA5 in the First Recipient Field.

6.3.4   A suggestion for implementation

A UA that requests *notifications* will need to have some mechanism to correlate the received EDINs with the expected EDINs. Since EDINs will be received from each *recipient* of each EDIM for which *notifications* are requested, the UA should maintain a table (or list) of the form shown in Table 6.2.

Table 6.2 - Expected EDINs

EDIM Identifier	Recipient ORName	type of EDIN requested and time expected

When an EDIN is received, its Subject EDIM field will contain an EDIM Identifier corresponding to a table entry, and its First Receiver Field will contain a Recipient ORName corresponding to a table entry.

Taken together, these two values will determine a unique entry in the table, and the UA will be able to determine whether the requested EDIN has been received.

As indicated, the UA should maintain, for each table entry, the latest time at which it expects the EDIN. If no EDIN is received within the stipulated time, the UA should assume that the EDIM has not reached its destination, and should take appropriate action.

> NOTE: If the same ORName was specified for several *recipients*, and *notifications* were requested from each, then the *originator* will receive indistinguishable EDINs, since the EDINs received will have the same Subject EDIM and the same First Recipient.
>
> If it is absolutely necessary to specify the same *recipient* ORAddress more than once for the same EDIM, then uniqueness can be achieved either by using different values in the Organizational Units *fields* of the ORName, or by using different Directory Names, which have the same ORAddress component (but specify different ORNames).

NOTE: Two PN EDINs may be received for the same EDIM, in the following case:

A UA may receive the same EDIM twice, due to re-transmission in case of suspected network problems.

As specified in clause 17.3 of X.435, a receiving UA may not be able to detect the reception of a duplicate EDIM. In this case, if *notifications* are requested, the UA will send a PN EDIN for the duplicate, and the original *originator* will receive two PN EDINs for the same EDIM.

7 Applying P_{edi} to possible scenarios

We are now in a position to understand how the several features of the P_{edi} protocol can be used in order to implement an X.400 data transmission mechanism for EDI in the different scenarios described in Chapter 2.

The reader should refer back to Chapter 2 for a description of the five different scenarios that we will consider.

7.1 The UA serves one application

In this simple scenario there is a one-to-one relationship between the UA and the application. Thus, the UA will never *forward* an EDIM, and the UA will be able to accept or refuse *responsibility* for the EDIM. PN or NN *notifications* will be sent by the UA.

The features provided by EDINs are used (including the security features that will be discussed in Chapter 10), but the *forwarding* mechanisms of P_{edi} are not used (and need not be implemented).

Note that a company that adopts this scenario does not need to acquire a separate copy of the UA software for each different application.

Most of the available software packages that provide implementations of the functions of a UA provide these functions for arbitrarily many distinct ORNames. That is, one physical copy of the UA software provides several logical UAs, each with its own distinct ORName.

Thus, one copy of the UA software can be configured to provide the required one-to-one relationship to each distinct application.

7.2 The UA is a corporate gateway that accepts responsibility

In this scenario a single UA (or a limited number of UAs) will receive (and send) all EDI Interchanges for the corporation. This UA will itself accept or refuse *responsibility* before routing the EDI Interchange to an application.

FN *notifications* will never be sent by the corporate gateway UA.

All PN and NN *notifications* will be sent by the corporate gateway UA.

Security features are provided between the external *originator* or *recipient* and the corporate gateway UA, and <u>not</u> between the external *originator* or *recipient* and the final *recipient* or *originator*.

If X.400 is not used for internal transmission within the corporation, then the *forwarding* mechanisms of P_{edi} are not used (<u>and need not be implemented</u>), and this scenario closely resembles the one described in section 7.1.

If X.400 is used for internal transmission within the corporation, then the *forwarding* mechanisms of P_{edi} are used as described in section 5.3, since *responsibility* is <u>not</u> *forwarded* even though the EDIM is *forwarded.*

If EDIMs are always *forwarded* unchanged, then the corporate gateway UA does <u>not</u> need to implement the place holder mechanism that allows body parts to be dropped when forwarding (section 3.5).

However, the ability to *forward* separately the several *body parts* of an EDIM is in general quite interesting in this scenario, and it is likely that most implementations of this scenario will have the ability to *forward* with changes.

For example, the corporate gateway UA could receive an EDIM containing both a purchase order and an engineering drawing specifying the part to be purchased. The *body part* containing the EDI Interchange could be *forwarded* to a purchasing system running on one computer, while the *body part* containing the drawing could be *forwarded* to a design system running on some other computer. (See also the discussion on cross-referencing in Chapter 8.)

Normally, there would a one-to-one relationship between applications and UAs within the corporation, so that within the corporation the situation described in section 7.1 would apply. Internal corporate UAs could therefore be simpler than the corporate gateway UA, since they might be incapable of *forwarding* EDIMs.

7.3   The UA is a corporate gateway that forwards responsibility

This scenario requires the full features of P_{edi}, and requires the corporation to use X.400 for internal transmission of EDI Interchanges.

In this scenario, security features can be provided end-to-end, from original *originator* to final *recipient*, even though a corporate gateway performs an intermediate routing function.

FN *notifications* will be sent by the corporate gateway UA. All PN and NN *notifications* will be sent by the final *recipient* UA. The corporate gateway UA will perform *forwarding* as described in section 5.2.

Since EDIMs are always *forwarded* unchanged, the corporate gateway UA does <u>not</u> need to implement the place holder mechanism that allows body parts to be dropped when forwarding (section 3.5).

Normally, there would a one-to-one relationship between applications and UAs within the corporation, so that within the corporation the situation described in section 7.1 would apply. Internal corporate UAs could therefore be simpler than the corporate gateway UA, since they might be incapable of *forwarding* EDIMs.

7.4   The UA is a clearing house

A VAN service can implement clearing house functions either in an MTA that it provides, or in an EDI UA that it provides, or in a combination of the UA and the MTA.

A full discussion of the possible options is beyond the scope of this book, due to the great variety of possible clearing house functions, and the many possible ways of implementing them.

However, a few general observations are relevant.

7.4.1   Using the MTA

In this scenario clearing house functions (such as mail-box storage, syntax conversion, selective routing, audit trails etc.) are provided by proprietary code that is embedded within the MTA operated by the VAN service.

The MTA itself must of course conform to the X.400 Recommendations, but this does not prevent it from performing additional functions that are beyond the scope of the X.400 Recommendations. (See for example section 15.3.)

7.4.2   Using the UA

The scenario described in section 7.2 corresponds to an in-house, corporate clearing house service.

A VAN service may choose to implement a full-feature EDI UA, and to use this UA to provide clearing house function. The full-feature EDI UA is capable of mail-box storage and selective routing, including the capability of routing separately the separate *body parts* of a single EDIM.

If the UA accepts *responsibility*, it can perform syntax conversion, and other tasks that require modification of the *body parts*. In addition, the UA can perform conversion to proprietary telecommunications protocols, in order to permit interworking with users who have not implemented X.400.

When converting to proprietary telecommunications protocols, the UA has the option to accept *responsibility* immediately, or to wait until confirmation is received that the transmission using the proprietary protocol has been completed successfully.

7.5   The UA is a small company gateway

This scenario is very similar to that described in section 7.3. The key difference is that the small company will not operate an internal X.400 network, but will rely on a public (PTT) or private (VAN) service to provide the X.400 network used to *forward* EDIMs. The small company will operate a single UA, and no MTAs.

Security features are provided end-to-end, from original *originator* to final *recipient*, even though the small company gateway performs an intermediate routing function.

FN *notifications* will be sent by the small company gateway UA.

All PN and NN *notifications* will be sent by the final *recipient* UA, which will be the small company gateway UA for some EDIMs, and some other external UA for other EDIMs.

The small company gateway UA will perform *forwarding* as described in section 5.2.

Since EDIMs are always *forwarded* unchanged, the small company gateway UA does <u>not</u> need to implement the place holder mechanism that allows body parts to be dropped when forwarding (section 3.5).

8 Cross-referencing

P_{edi} includes a rather generic mechanism for cross-referencing among the several *body parts* of an EDIM. The mechanism is nothing more than a *field* in the EDIM Heading, which can be used to associate arbitrary application-defined reference numbers with specific *body parts*.

The *field* is the *Cross Referencing Information field* of the EDIM Heading, defined in clause 8.2.12 of X.435. This *field* is a table, which takes the form of Table 8.1.

Table 8.1 - Cross-referencing

Application Cross Reference	Message Reference	Body Part Reference

The Application Cross Reference is an arbitrary binary string, specified by the EDI application itself. For example, it could be a reference number contained in an EDIFACT RFF *segment*.

The Message Reference, which is optional, is used to hold the EDIM Identifier, or IPM Identifier, of the X.400 *message* to which the Body Part Reference refers. This *field* is used only if the Body Part Reference refers to an X.400 *message* other than the EDIM in which the Cross Referencing Information appears.

The Body Part Reference is simply an integer that identifies a particular *body part* within a particular *message*.

Figure 8.1 shows an example:

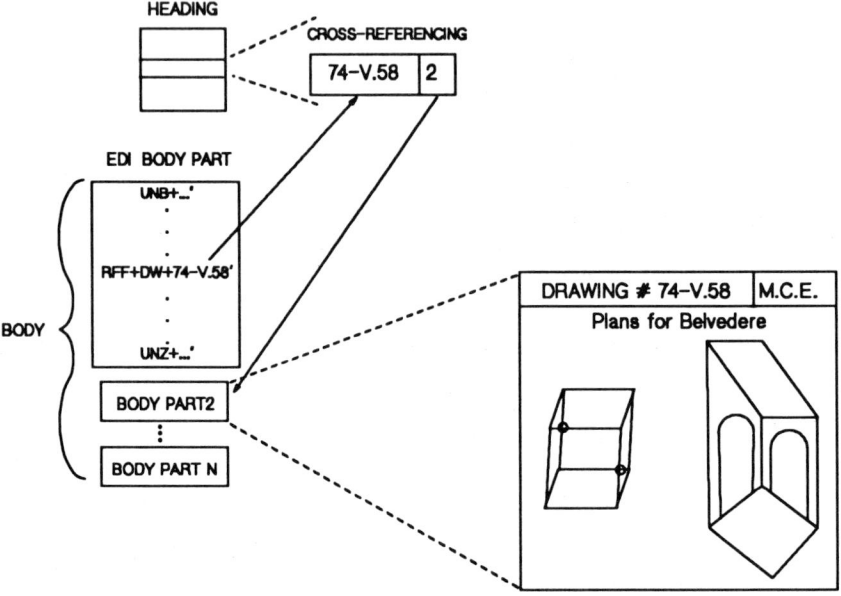

Figure 8.1 - Cross-referencing

The EDI Interchange contains a reference to a drawing. The drawing number is "74-V.58", and this reference number appears in an RFF *segment* within the EDI Interchange. The Cross Referencing *field* of the EDIM Heading holds this same reference number, and the associated Body Part Reference, which is just the number "2", corresponding to the fact that the drawing is the second *body part* in the EDIM.

Body Part 2 contains the actual drawing which, in this example, also contains the reference number "74-V.58".

The Recommendations F.435 and X.435 do <u>not</u> specify how the Cross Referencing *field* of the EDIM Heading should be used. Annex E of F.435 contains an example similar to that of Figure 8.1. However, implementors are free to use the Cross Referencing *field* as they wish, provided, of course, that they comply with the syntax specified in X.435.

When creating an EDIM, the Cross Referencing *field* can be created only by a sophisticated EDI User Agent that is capable of accepting from the application all the data needed to create the *field.*

The following steps could provide a starting point when planning an implementation of the cross-referencing features:

1. EDI applications create the contents of the Application Cross Reference *fields* (ACRs).

2A. If the cross-referenced *body part* is within the EDIM, the EDI application passes to the UA the ACRs, the EDI Interchange and the other *body parts* to which the ACRs refer. The EDI UA creates the EDIM and the Cross Reference *field* in the EDIM Heading.

2B. If cross-referenced *body part* is not within the EDIM, the EDI application passes to the UA the ACRs, the EDI Interchange, the Body Part References and the Message References of the *body parts* (contained in other X.400 *messages*) to which the ACRs refer. The EDI UA creates the EDIM and the Cross Reference *field* in the EDIM Heading.

Cases 2A and 2B are not mutually exclusive.

3. The EDI UA returns to the EDI application (for logging and reference purposes) the EDIM Identifier and the cross-reference data.

4. The EDI UA *submits* the EDIM to the MTA.

When receiving an EDIM, the Cross Referencing data contained in the EDIM Heading could be used by a sophisticated EDI User Agent in order to *forward* separately, to different *recipients*, the *body parts* containing the EDI Interchange and the referred-to *body parts* (for example, drawings).

An EDI User Agent or application process wishing to correlate a *body part* with a reference number found inside an EDI Interchange uses the reference number to perform a lookup in the Cross Referencing *field*. It finds the corresponding *body part* reference in the Cross Referencing *field*, and this can then be used to locate and extract that *body part*.

As we have seen in section 3.5, *body parts* may be dropped when *forwarding*, so that the EDI User Agent may *forward* an EDI Interchange after dropping some *body parts*, and, separately, and to a different *recipient*, it may *forward* a drawing after dropping the EDI Interchange *body part*.

Note that the entire EDIM Heading is always included in any *forwarded* EDIM, so that the Cross Referencing *field* is available to all recipients, even if they do not receive all the *body parts*.

Specification of the logic required by the EDI User Agent in order to perform this type of *body part* dropping and sophisticated forwarding is beyond the scope of the Recommendations F.435 and X.435, and will not be discussed further here.

9   Message Store

X.413 defines the Message Store (MS), an <u>optional</u> facility designed to provide X.400 messaging services to users who do not operate an MTA.

The Message Store provides storage for X.400 *messages* that cannot be delivered immediately from an MTA to a UA, for example, because the UA operates on a PC that is turned off from time to time.

When an MS is present, the UA is connected to the MS, and the MS to the MTA, as illustrated in Figure 3.1 of Chapter 3. If present, an MS must obey the same operating rules as a UA.

In addition to the general MS features specified in X.413, clause 18 of X.435 specifies EDI-specific MS features. These features are designed to allow small users to benefit fully from P_{edi}, even if they do not operate an MTA.

The basic features of the P_{edi} MS are:

-    maintaining EDI-specific MS Attributes

-    EDI forwarding with responsibility not accepted Auto-Action

-    EDI forwarding with responsibility accepted Auto-Action

9.1   EDI-Specific MS Attributes

When an EDIM or an EDIN is kept in the Message Store, the MS transforms the Heading *field* of the EDIM, or most *fields* of the EDIN, into *MS Attributes*, that can be used to specify retrieval of *messages*.

For example, the EDI Message Type *field* of the EDIM Heading becomes an MS Attribute, and the UA can request retrieval of EDIMs on the basis of the value of this MS Attribute (for example, it can select for retrieval all EDIMs with the value "INVOIC" in the MS Attribute corresponding to the EDI Message Type).

Clause 18.7 of X.435 defines in detail the MS Attributes.

The manner in which the MS Attributes are derived from an EDIM is illustrated schematically in Figure 9.1.

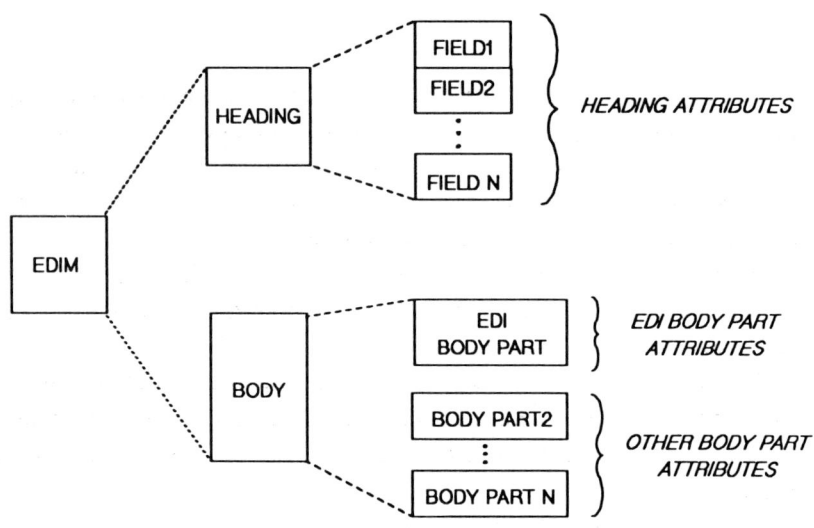

Figure 9.1 - Message Store Attributes

That is, the values contained in the *fields* of the EDIM are mapped into MS Attributes. The most important MS Attributes for an EDIM are:

Heading: the entire EDIM Heading becomes an Attribute.

Heading Fields: each of the EDIM Heading *fields* becomes an Attribute.

Recipient Sub-Field: each of the sub-*fields* of the Recipient *field* becomes an Attribute.

Body: the entire EDIM Body becomes an Attribute.

Body Analyses: the length in octets (bytes) of the EDI Interchange becomes an Attribute.

Example 9.1 shows the MS Attributes for the EDIM shown in Example 3.1 of section 3.3.

Example 9.1 - MS Attributes

MS Attributes	
Field	Value
This EDIM	CH/ARCOM/HP/EDI/PURCH, 90/01/26/10:21:04
Originator	CH/ARCOM/HP/EDI/PURCH
Recipients	
Recipient	CH/ARCOM/SUPPLIER
Notification Requests	NN
Interchange Recipient	SUPPLIER
Interchange Control Reference	135
EDI Body Part Type	EDIFACT, ISO 646
EDI Message Type	ORDERS
Interchange Sender	HP

X.413 specifies the concept of an *Auto-Action*. This is an action which the MS performs automatically on behalf of the UA. The UA must of course instruct the MS to perform specific Auto-Actions for a specific set of *messages*.

The UA can specify a selection criterion based on the MS Attributes, and instruct the MS to perform a certain Auto-Action for all *messages* that satisfy the selection criterion.

For example, in the IPM protocol, a UA can instruct the MS to Auto-Forward all *messages* that come from a certain *originator*.

The Auto-Actions defined in X.413 are not generally useful for P_{edi}, therefore X.435 defines the EDI-specific Auto-Actions "forwarding with responsibility not accepted" and "forwarding with responsibility accepted". These Auto-Actions are explained in sections 9.2 and 9.3.

In both cases, the MS must obey the same operating rules as a UA that implements *forwarding* (see section 5.4 and in particular sub-section 5.4.4).

Requests for security features may restrict the availability of certain MS features. See section 10.7.

9.2   Forwarding with responsibility not accepted Auto-Action

This Auto-Action is defined in clause 18.6.1 of X.435 It is identical to the *forwarding* case discussed in section 5.2.

When the UA instructs the MS to perform this Auto-Action, it instructs the UA to always forward *EDI Responsibility* for an EDIM that it forwards.

That is, the UA instructs the MS to forward all EDIMs that satisfy certain selection criteria based on the MS Attributes, and, in addition, it instructs the MS to generate FN EDINs for the forwarded EDIMs, if any are requested, and to forward the Notification Requests in the EDIM to the *recipient* of the forwarded EDIM.

The MS always forwards the EDIM unchanged, that is, no body parts are added or dropped, and it forwards the EDIM to at most one *recipient*.

For example, a small-company UA could instruct a commercial Value Added Network service MS to EDI Forward with FN all EDIMs with Message Type "INVOIC" to an external accounting service that would accept *EDI Responsibility* and pay the invoices on behalf of the small company (see Chapter 6 for a discussion of *EDI Responsibility*).

9.3   Forwarding with responsibility accepted Auto-Action

This Auto-Action is defined in clause 18.6.2 of X.435. It is identical to the *forwarding* case discussed in section 5.3.

When the UA instructs the MS to perform this Auto-Action, it instructs the UA to always accept *EDI Responsibility* for an EDIM that it forwards (see Chapter 6 for a discussion of *EDI Responsibility*).

That is, the UA instructs the MS to forward all EDIMs that satisfy certain selection criteria based on the MS Attributes, and, in addition, it instructs the MS to generate PN EDINs for the forwarded EDIMs, if any are requested.

If no EDINs are requested, the MS can still Auto-Forward the EDIM, and to do so it uses the rules for the "forwarding with responsibility accepted" Auto Action.

The MS always forwards the EDIM unchanged, that is, no body parts are added or dropped. The EDIM may be forwarded to more than one *recipient*.

For example, a small-company UA could instruct a commercial Value Added Network service MS to EDI Forward with PN all EDIMs with Message Type "ORDERS" to both an EDI UA operating within the small company, and to an external accounting service.

9.4 Registration of Auto-Actions

If an Auto-Action is desired, it must be *registered* (see X.413), and the selection criteria for the Auto-Action must be specified.

Since very general selection criteria can be specified for each *registered* Auto-Action, it can happen that more than one Auto-Action is requested for a particular EDIM. Clause 18.8.1 of X.435 specifies the priority rules that are used in order to decide which Auto-Action to perform if several are requested. The rules can be summarized as follows:

1) Look for "forwarding with responsibility accepted" Auto-Actions, and perform as requested. Several Auto-Actions can be performed by the same MS for the same EDIM, resulting in multiple *forwarding*.

2) If no "forwarding with responsibility accepted" Auto-Actions are requested for the EDIM, look for a "forwarding with responsibility not accepted" Auto-Action, and perform as requested. At most one of these Auto-Actions can be performed by the same MS for the same EDIM, so in this case multiple *forwarding* cannot take place.

10 Security features

The P_{edi} protocol includes a number of security features that are unique to it, and that allow end-to-end security services to be provided, even when EDIMs are forwarded.

The fundamental security requirements can be summarized with a few key words: <u>authentication</u>, <u>integrity</u> and <u>confidentiality</u>.

Availability, often mentioned when security is discussed, is not, properly speaking, a security issue, although security mechanisms may help to detect network unavailability.

We begin with a review of EDI user security requirements. The reader may also wish to review Annex C of F.435 and Annexes A and F of X.509. Annex I of X.435 specifies enhancements to the security model defined in X.402.

10.1   EDI user security requirements

In early 1989, as part of the TEDIS program, Directorate General XIII of the Commission of European Communities made publicly available the results of an extensive study on the security requirements of EDI users. The study was conducted on behalf of the Commission by ICL.

We summarize here the principal findings of this study.

10.1.1   Threats

EDI users ranked threats to security, in order of priority, as follows:

1.   Loss of service.

2.   Disclosure of information.

3.   Unauthorized network access by insiders.

4.   Fraud.

5.   Unauthorized network access by outsiders.

10.1.2   Required Facilities

The following security facilities or features were felt to be required in order to protect against the threats listed above. The facilities were rated as "essential" or "important".

As discussed later in this chapter, P_{edi} provides many of the required facilities.

10.1.3   Essential

A.   User authentication.

B.   Message integrity.

C.   Confirmation of end-to-end delivery.

10.1.4   Important

D.   Message confidentiality.

E.   Network service operational security.

F.   Auditability.

G.   Network service harmonized security levels.

H.   Non-repudiable confirmation of receipt.

10.1.5   Threats vs. facilities

We can make a specific list showing which security facilities can be used to protect against which threats:

1. Loss of service: This is not correctly speaking a security problem, since service can be lost even when security is not compromised.

However, confirmation of end-to-end delivery (A) does prove that service has not been lost, and message integrity (B) does ensure that an EDI Interchange is not corrupted during transmission.

2. Disclosure: Message confidentiality (D) provides protection against disclosure.

3. Unauthorized access by insiders: Network service operational security (E) protects against this threat. Auditability (F) provides additional protection.

4. Fraud: User authentication (A), confirmation of end-to-end delivery (C) and non-repudiable confirmation of receipt (H) are the facilities required to prevent fraud. Auditability (F) provides additional protection.

5. Unauthorized access by outsiders: Network service operational security (E) and network service harmonized security levels (G) protect against this threat. Auditability (F) provides additional protection.

10.2   How P$_{edi}$ provides the required facilities

NOTE: Throughout this chapter, it is assumed that the boundary between the EDI User Agent and the EDI application is <u>trusted</u>. That is, it is assumed that there is no need for security facilities when transmitting data between the UA and the application.

Thus, it is assumed that when security is provided from the *originator* UA to the *recipient* UA, it is also provided from the originating EDI application to the receiving EDI application.

This assumption may be invalid for some of the scenarios discussed in Chapters 2 and 7. In this case, appropriate UA to EDI application security facilities need to be provided. Provision of such facilities in general is beyond the scope of X.435, and will not be discussed further, with the exception of the EDI Application Security Element, discussed in section 10.6.

10.2.A   Authentication

The purpose of this security facility is to be able to prove that *message* does indeed come from the *originator* whose ORName appears in the Originator *field* of the *message*.

This facility is equivalent to the familiar concept of a signature.

The P_{edi} protocol provides this facility by relying on the Origin Authentication features of the P1 *envelope*, as specified in 10.2.1.1.1 and 10.3.1.2.1 of X.402 and 8.2.1.1.1.29 of X.411.

A minimal knowledge of asymmetric public key encryption methods is required in order to understand how this facility is provided. The next four paragraphs provide a highly simplified introduction to these methods. A more detailed explanation is given in Annex B of X.509.

In most commonly used encryption methods, a single <u>secret</u> key is used by both originator and recipient. The originator uses the key to encrypt the message and the recipient uses the same key to decrypt the message. Some secure method is used to convey the secret key from the originator to the recipient. If the secret key becomes known to others while it is being conveyed, then the encryption is no longer secure, since any one who has the key can decrypt the message.

Since the recipient can encrypt as well as decrypt, single (also called symmetric) key encryption methods cannot be used to provide authentication.

Authentication can be provided only when the originator uses a secret key known to him and to him <u>only</u>. Since no one else knows the originator's key, no one else can encrypt messages with that key.

In order to avoid the burden of having numerous pairs of encrypting and decrypting secret keys, it would be desirable to use <u>secret</u> keys for encryption, and <u>public</u> keys for decryption. That is, each originator creates a matched pair of encryption and decryption keys. The encryption key is kept secret, the decryption key is made public. Encryption methods that allow for secret encryption and public decryption keys are called asymmetric public key encryption algorithms. One specific algorithm of this type is described in Annex C of X.509 (the well-known RSA algorithm).

If asymmetric public key encryption is available, the following procedures are used by the *originator* and the *recipient*.

Originator:

a)   Apply a hash function to the *content* of the X.400 *message*.

NOTE: A *hash function* is a computation that reduces a large number of bits to a smaller number of bits, in such a way that all of the original bits influence the outcome of the computation. See Annex D of X.509.

---

b) Encrypt the result of the hash function using asymmetric public key encryption, and using the *originator's* <u>secret</u> key.

c) Send the encrypted result of the hash function together with the *message*.

Recipient:

d) Apply the hash function to the *content* of the X.400 *message*.

e) Decrypt the encrypted result of the hash function which was sent together with the *message* (step c) above). The decryption uses the <u>public</u> key of the *originator*.

f) Compare the results of steps d) and e). If the decrypted result of the hash function is identical to the result of d), then the *message* was indeed sent by the *originator*, since no one else could have encrypted the result of the hash function with the *originator's* <u>secret</u> key.

Figure 10.1 illustrates this process:

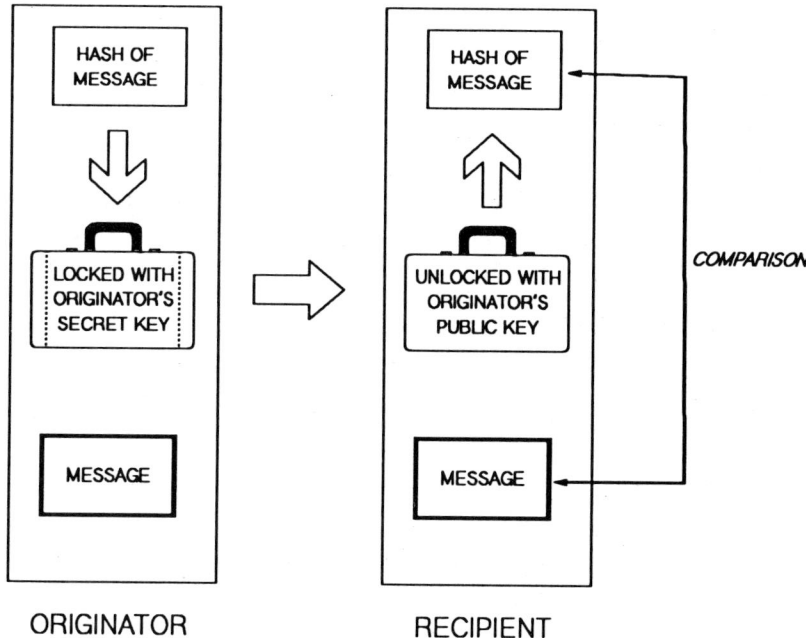

Figure 10.1 - Authentication

10.2.B   Message integrity

The purpose of this security facility is to be able to prove that the *message submitted* by the *originator* was not changed in any way before it was received by the *recipient*.

This facility is similar to the familiar concept of a sealed envelope.

Again, the P_{edi} protocol relies on the basic security features provided by X.400. Message integrity can be achieved in two different ways:

-    with asymmetric public key encryption (as explained above);

-    with symmetric secret key encryption (see 8.2.1.1.1.28 of X.411).

When asymmetric public key encryption is used, the following steps are performed:

Originator:

a)   Apply a hash function to the *content* of the X.400 *message*.

b)   Encrypt the result of the hash function using asymmetric public key encryption, and using the *originator's* secret key.

c)   Send the encrypted result of the hash function together with the *message*.

Recipient:

d)   Apply the hash function to the *content* of the X.400 *message*.

e)   Decrypt the encrypted result of the hash function which was sent together with the *message* [step c) above].   The decryption uses the public key of the *originator*.

f)   Compare the results of steps d) and e). If the decrypted result of the hash function is identical to the result of d), then the *message* sent by the *originator* has indeed not been changed in any way, since no one but the *originator* could have created the encrypted result of the hash function.

Figure 10.1 illustrates this process.

When symmetric secret key encryption is used, the following steps are performed:

Originator:

g) Apply a hash function to the *content* of the X.400 *message*.

h) Encrypt the result of the hash function, using the <u>secret</u> key known to both the *originator* and the *recipient*.

i) Send the encrypted result of the hash function together with the *message*.

Recipient:

j) Apply the hash function to the *content* of the X.400 *message*.

k) Decrypt the encrypted result of the hash function which was sent together with the *message* [step i) above].

l) Compare the results of steps j) and k). If the decrypted result of the hash function is identical to the result of j), then the *message* sent by the *originator* has indeed not been changed in any way, since no one but the *originator* could have created the encrypted result of the hash function.

Note that when symmetric secret key encryption is used, there must be some secure way for the *recipient* and the *originator* to exchange the secret key. Provision of such a secure method for exchanging keys is beyond the scope of the X.400 standards.

Once *recipient* and *originator* posses a secret key, and if they are certain that the secret key has not been compromised, they can exchange new secret keys by sending the new keys within *messages* encrypted using the old key. Greater security can be achieved if some process other than X.400 is used for exchanging new keys.

NOTE: When either of the above processes is used, message integrity is guaranteed if and only if no two messages will yield the same result of the hash function. It is obvious that the only hash function with this property is the hash function that returns the *content* unchanged.

In practice, a very high level of confidence in message integrity can be achieved with hash functions that return a result much shorter than the *content*. Specification of actual hash functions is beyond the scope of the X.400 standards. One suitable hash function is defined in Annex D of X.509.

10.2.C   Confirmation of end-to-end delivery

The purpose of this security facility is to be able to prove that the *message submitted* by the *originator* was indeed received by the *recipient*.

This facility is similar to the familiar concept of a signed return receipt.

The EDI Notifications discussed in Chapters 4 and 6 provide the mechanism required to implement this facility. As discussed, EDI Notifications are end-to-end notifications.

The following steps are used to provide confirmation of delivery:

Originator of EDIM:

a)   Request PN and NN EDIN, with authentication. This is done by setting the Notification Security *field* of the EDI Notification Request *field* to the value "proof" (see section 10.4).

Recipient of EDIM, Originator of EDIN:

b)   If the received EDIM requests PN or NN with "proof", use the authentication procedure outlined in a), b) and c) of sub-section 10.2.A in order to generate the EDIN.

That is, generate an authenticated EDIN.

Originator of EDIM, Recipient of EDIN:

c)   When the EDIN is received, use the verification procedures outlined in d), e) and f) of sub-section 10.2.A in order to prove that the *originator* of the EDIN is indeed the *originator* indicated by the EDIN Originator *field* of the EDIN.

10.2.D   Message confidentiality

The purpose of this security facility is prevent anyone other than the *recipient* from reading the *contents* of the *message*.

This facility is similar to the familiar concept of a lined envelope (that is, an envelope that is sufficiently opaque so that the contents of the envelope cannot be read).

The P_{edi} protocol provides this facility by relying on the *message* encryption features of the P1 *envelope*, as specified in 10.2.3.2 and 10.3.3.1 of X.402. The standard does not mandate any particular encryption method. The method used is identified as specified in 8.2.1.1.1.27 of X.411.

10.2.E   Network service operational security

The purpose of this security facility is to prevent tampering with the network, and to prevent eavesdropping.

This facility is similar to the familiar concept of security measures taken by post offices (locked doors, security guards, etc.)

Such security features are an integral part of the X.400 protocols as a whole, and of the related lower-level protocols such as X.25, and will not be discussed in detail here.

10.2.F   Auditability

The purpose of this security facility is to be able to verify what has passed through the network, so that confidence in the effectiveness of network security measures can be established.

This facility is similar to the familiar concept of an audit.

It would appear that creating and maintaining log files is essential to provision of auditability.

Creation of log files is considered a local matter, beyond the scope of the X.400 standards. Thus, considerable work remains to be done in this important area.

Auditable log files could be provided by Value Added Network services, or by specialized certified agencies.

10.2.G   Network service harmonized security levels

The purpose of this security facility is to ensure consistency of security levels, so that users of the network can be confident that security is maintained consistently throughout the network.

This facility is similar to the familiar concepts of world-wide standards for handling of registered mail, or company-wide definitions of "company confidential", etc.

Harmonized security services, as specified in X.402 and X.411, are an integral part of the X.400 protocol, as we have seen in this section.

10.2.H   Non-repudiable confirmation of receipt

The purpose of this security facility is to obtain non-repudiable proof that the *message submitted* by the *originator* was indeed received by the *recipient*.

This facility is similar to what would be provided by requesting a notarized signature on the familiar return receipt.

The EDI Notifications discussed in Chapters 4 and 6 provide the mechanism required to implement this facility. As discussed, EDI Notifications are end-to-end notifications.

The following steps are used to provide non-repudiable confirmation of receipt:

Originator of EDIM:

a) Request PN and NN EDIN, with non-repudiable authentication. This is done by setting the Notification Security *field* of the EDI Notification Request *field* to the value "non-repudiation" (see section 10.4).

Recipient of EDIM, Originator of EDIN:

b) If the received EDIM requests PN or NN with "non-repudiation", use the authentication procedure outlined in a), b) and c) of sub-section 10.2.A in order to generate the EDIN, and, in addition obtain a "non-repudiable certificate of origin", which is sent together with the EDIN.

Originator of EDIM, Recipient of EDIN:

c) When the EDIN is received, use the verification procedures outlined in d), e) and f) of sub-section 10.2.A in order to prove that the *originator* of the EDIN is indeed the *originator* indicated by the EDIN Originator *field* of the EDIN. The presence of the "non-repudiable certificate of origin" provides the requested non-repudiability.

NOTE: The X.400 standards do not specify how to obtain a "non-repudiable certificate of origin". There are several methods for doing this, some of which are conceptually similar to the familiar process of obtaining a notarized signature.

That is, a known signature (the notary's) is used to prove that the *originator's* signature is indeed authentic.

The only difference between "non-repudiable confirmation of receipt" and "confirmation of end-to-end delivery" is the use of the "non-repudiable certificate of origin" to provide the equivalent of a notarizing function.

10.3 How P_{edi} provides additional facilities

The P_{edi} protocol includes mechanisms that allow provision of security facilities in addition to those examined above. The additional facilities provided are:

I. Proof of content received.

J. Non-repudiable proof of content received.

10.3.I Proof of Content Received

The purpose of this security facility is for the *originator* to be certain that the *message* received by the *recipient* was indeed the one sent.

This facility is similar to what would be provided if the recipient of a letter sent back to the originator a photo-copy of the letter received.

The EDI Notifications discussed in Chapters 4 and 6 provide the mechanism required to implement this facility. As discussed, EDI Notifications are end-to-end notifications.

The following steps are used to provide proof of content received:

Originator of EDIM:

a) Request PN and NN EDIN, with authentication. This is done by setting the Notification Security *field* of the EDI Notification Request *field* to the value "proof" (see section 10.4).

In addition, apply a hash function to the *content* of the *message*, encrypt the result of the hash function, and transmit it together with the *message*.

Recipient of EDIM, Originator of EDIN:

b)   If the received EDIM requests PN or NN with "proof", use the authentication procedure outlined in a), b) and c) of sub-section 10.2.A in order to generate the EDIN.

In addition, place the encrypted result of the hash function generated in a) above in the Original Content Integrity Check *field* of the EDIN (see section 10.5).

Originator of EDIM, Recipient of EDIN:

c)   When the EDIN is received, use the verification procedures outlined in d), e) and f) of sub-section 10.2.A in order to prove that the *originator* of the EDIN is indeed the *originator* indicated by the EDIN Originator *field* of the EDIN.

In addition, verify that the encrypted result of the hash function generated in a) above matches the value contained in the Original Content Integrity Check *field* of the EDIN.

Figure 10.2 illustrates this process.

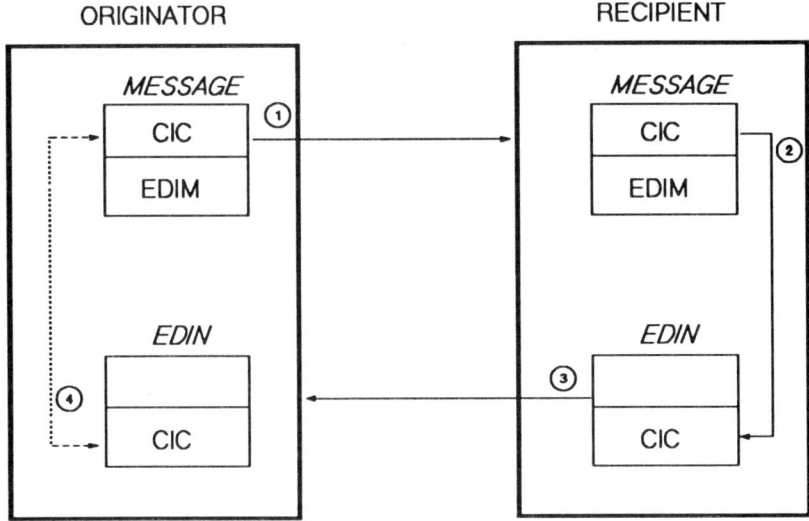

CIC = Content Integrity Check

Figure 10.2 - Proof of Content

The steps illustrated Figure 10.2 are:

1. The *originator* applies a hash function to the *content* of the *message*, encrypts the result of the hash function, and transmits it together with the *message*, in the Content Integrity Check *field* (CIC).

2. The *recipient* copies the CIC from the received EDIM into the EDIN.

3. The *recipient* transmits the EDIN to the *originator*.

4. The *originator* compares the CIC received in the EDIN with the CIC transmitted with the original EDIM.

NOTE: If the *originator* of the EDIM does not transmit an encrypted result of a hash function together with the *message*, proof of content received can still be provided, if the *recipient* of the EDIM places the entire received EDIM in the Original Content *field* of the EDIN (see section 10.5).

10.3.J   Non-repudiable proof of content received

The purpose of this security facility is for the *originator* to be certain that the *recipient* cannot claim that he did not receive the *message* that was sent.

This facility is similar to the familiar concept of returning to the originator a signed and notarized copy of a contract that had been previously signed and notarized by the originator.

The EDI Notifications discussed in Chapters 4 and 6 provide the mechanism required to implement this facility.   As discussed, EDI Notifications are end-to-end notifications.

The following steps are used to provide non-repudiable proof of content received:

Originator of EDIM:

a)   Request PN and NN EDIN, with non-repudiable authentication.   This is done by setting the Notification Security *field* of the EDI Notification Request *field* to the value "non-repudiation" (see section 10.4).

In addition, apply a hash function to the *content* of the *message*, encrypt the result of the hash function, and transmit it together with the *message*.

Recipient of EDIM, Originator of EDIN:

b)   If the received EDIM requests PN or NN with "non-repudiation", use the authentication procedure outlined in a), b) and c) of sub-section 10.2.A in order to generate the EDIN, and, in addition, obtain a "non-repudiable certificate of origin", which is sent together with the EDIN (see also sub-section 10.2.H).

In addition, place the encrypted result of the hash function generated in a) above in the Original Content Integrity Check *field* of the EDIN (see section 10.5).

Originator of EDIM, Recipient of EDIN:

c)   When the EDIN is received, use the verification procedures outlined in d), e) and f) of sub-section 10.2.A in order to prove that the *originator* of the EDIN is indeed the *originator* indicated by the EDIN Originator *field* of the EDIN. The presence of the "non-repudiable certificate of origin" provides the requested non-repudiability.

In addition, verify that the encrypted result of the hash function generated in a) above matches the value contained in the Original Content Integrity Check *field* of the EDIN.

NOTE: If the *originator* of the EDIM does not transmit an encrypted result of a hash function together with the *message,* proof of content received is still provided, because the *recipient* of the EDIM places the entire received EDIM in the Original Content *field* of the EDIN (see section 10.5).

10.4 EDI Notification and Reception Security Requests

Clause 8.2.3.3 of X.435 specifies the following security requests, which can be made of any of the *recipients* of an EDIM:

-   Notification Security

-   Reception Security

These requests correspond to *fields* in the EDIM Heading. Either *field* can have the value "proof" or "non-repudiation". That is, the *originator* of the EDIM can request Notification Security "proof", or Notification Security "non-repudiation", or Reception Security "proof", or Reception Security "non-repudiation".

The only difference between "proof" and "non-repudiation" is the use of the "non-repudiable certificate of origin" to provide the equivalent of a notarizing function (see also sub-section 10.2.H).

10.4.1 Notification Security

The Notification Security *field* is used to request that any requested EDINs be authenticated, as explained in sub-section 10.2.C.

The security request applies to PNs, NNs and FNs.

As specified in clause 17.1.3 of X.435, a *recipient* UA must comply with the security requests, and generate authenticated EDINs as requested.

10.4.2 Reception Security

The Reception Security *field* is used to request that proof of content received be returned in a requested EDIN, as explained in sub-section 10.3.I.

The security request applies to PNs, NNs and FNs.

As specified in clause 17.1.3 of X.435, a *recipient* UA must comply with the security requests, and place the proof of content in the appropriate *field* of the EDIN (see section 10.5).

Example 10.1 shows what the example EDIM shown in Example 3.1 of section 3.3 would look like if both Notification Security and Reception Security were requested.

Example 10.1 - Security Requests

EDIM Heading	
Field	Value
This EDIM	CH/ARCOM/HP/EDI/PURCH, 90/01/26/10:21:04
Originator	CH/ARCOM/HP/EDI/PURCH
Recipients	
Recipient	CH/ARCOM/SUPPLIER
Notification Requests	PN
Notification Security	PROOF
Reception Security	PROOF
Interchange Recipient	SUPPLIER
Interchange Control Reference	135
EDI Body Part Type	EDIFACT, ISO 646
EDI Message Type	ORDERS
Interchange Sender	HP
EDIM Body	

```
    UNB+UNOA:1+HP+SUPPLIER+90/01/26+135'
    UNH+1+ORDERS:1'
    ...
    UNZ+1+135'
```

10.5 EDIN security elements

Clause 9.1.5 of X.435 specifies the following security elements, which are part of the Common Fields of the EDIN:

- Original Content

- Original Content Integrity Check

- EDI Application Security

Clause 17.1.3 of X.435 specifies how these *fields* are filled in when an EDIN is created.

EDI Application Security in discussed in section 10.6. The other *fields* are discussed in sub-sections 10.5.1 and 10.5.2.

In addition to the above, the Forwarded Reason Code *field* defined in clause 9.4.2 of X.435 includes the boolean *field* FNUAMS Security Check, which can be used for security purposes. The use of this *field* is discussed in sub-section 10.5.3

10.5.1 Original Content

Bullet b) 2) of 17.1.3 of X.435 specifies that the *content* of the received *message* is placed in this field, if Reception Security is requested, and the "content-integrity-check" field of the P1 *envelope* is <u>not</u> present.

10.5.2 Original Content Integrity Check

Bullet b) 1) of 17.1.3 of X.435 specifies that the "content-integrity-check" *field* of the received P1 *envelope* is placed in this field, if Reception Security is requested.

Example 10.2 shows what the EDIN would look like, in response to the example EDIM shown in Example 10.1 of section 10.4.

Example 10.2 - EDIN with Security Fields

EDIN Common Fields	
Field	Value
Subject EDIM	CH/ARCOM/HP/EDI/PURCH, 90/01/26/10:21:04
EDIN Originator	CH/ARCOM/SUPPLIER
First Recipient	CH/ARCOM/SUPPLIER
Notification Time	90/01/26/12:43:54
Security Elements Content Integrity Check	XX

NOTE: Because Notification Security was set to "proof" in the EDIM, the X.400 *message* containing the EDIN will be authenticated, as explained in sub-section 10.2.A.

The Content Integrity Check *field* is present in the EDIN because Reception Security was set to "proof" in the EDIM. Its value (shown as "XX" in the example) will be taken from the X.400 *message* containing the EDIM, as explained in sub-section 10.3.I.

10.5.3   FNUAMS Security Check

The Forwarded Reason Code *field* defined in clause 9.4.2 of X.435 includes the boolean *field* FNUAMS Security Check. How this field should be used is not specified in the standard. However, it can be used when forwarding with responsibility not accepted in order to convey the information that the security elements in the received EDIM (which has been forwarded) have been checked and found to be valid.

10.6   EDI Application Security

X.435 provides *fields* that can be used to provide end-to-end security between EDI applications, even if the boundary between the UA and the application cannot be trusted. These *fields* are:

- EDI Application Security Element in the EDIM Heading.

- EDI Application Security Element in the EDIN Common Fields.

The standard does not specify how to use these *fields*. One possible way to use the *fields* is shown below.

10.6.1 EDIM Heading

Clause 8.2.11 of X.435 defines the EDI Application Security *field* in the EDIM Heading.
    One way to use this field would be as follows, in order to provide authentication of the EDI application that created the EDI Interchange contained in the EDIM:

Sending EDI application:

a) Apply a hash function to the EDI Interchange.

b) Encrypt the result of the hash function using asymmetric public key encryption, and using the sender's <u>secret</u> key.

c) Place the encrypted result of the hash function in the EDI Security Elements *field* of the EDIM Heading.

Receiving EDI application:

d) Apply the hash function to the EDI Interchange.

e) Decrypt the contents of the EDI Security Elements *field* of the EDIM Heading of the received EDIM, using the <u>public</u> key of the sending EDI application.

f) Compare the results of steps d) and e). If the decrypted result of the hash function is identical to the result of d), then the EDI Interchange was indeed sent by the sender, since no one else could have encrypted the result of the hash function with the sender's <u>secret</u> key.

10.6.2 EDIN Common Fields

Clause 9.1.5 of X.435 defines the EDI Application Security *field* in the EDIN Common Fields. Clause 17.1.3 of X.435 does <u>not</u> contain any detailed instructions on the use of this *field*.

One way to use this field in order to provide end-to-end confirmation of delivery, with proof of contents received, between the sending and receiving EDI applications, would be as follows:

Sending EDI application:

g)   Apply a hash function to the EDI Interchange, and save the result for later use.

In addition, request the UA to request PN Notification.

Receiving EDI application:

h)   Apply the same hash function of g) above to the EDI Interchange.

Encrypt the result of the hash function, using asymmetric public key encryption, and the <u>secret</u> key of the receiving EDI application.

Request the UA to place this encrypted value in the EDI Security Elements *field* of the EDIN that will be generated in response to the request for PN Notification.

Sending EDI application:

i)   Obtain from the UA the value of the EDI Security Elements *field* of the EDIN received in response to the request for PN Notification.

Decrypt this value using asymmetric public key encryption and the <u>public</u> key of the receiving EDI application.

Compare the decrypted result with the saved result of the hash function of g) above. If the two are identical, then the EDI Interchange was indeed received by the EDI application to whom it was sent.

10.7   Security Restrictions

Security policies in force for certain implementations may restrict the availability of certain features of P_{edi} (for example, forwarding). In addition to locally defined restrictions, X.435 imposes the following restriction if security features are requested:

-       an MS cannot perform the "forwarding with responsibility
        accepted" Auto-Action (see clause 18.8.1 of X.435).

10.8   Security and Forwarding

Clause 8.2.5 of X.435 specifies that security elements may be checked when
forwarding with responsibility <u>not</u> accepted.   As noted in 10.5.3, if the
originator requested FN Notifications the FNUAMS Security Check *field* in
the EDIN can be used to inform the *originator* that security has been
checked.

In addition, clause 8.2.5 of X.435 specifies that the security elements shall
be checked when responsibility is accepted (that is, when a PN is sent, if PN
Notification was requested).

When forwarding with responsibility accepted takes place, the
Responsibility Forwarded *field* is not present.   Therefore, when scanning
through the nested structure of a *forwarded* EDIM, a UA can determine at
which point security was checked, by finding the level that does not have the
Responsibility Forwarded *field* set to "true".   See section 5.4 (in particular
sub-section 5.4.1) for rules related to the Responsibility Forwarded *field*.

11 Use of X.500 Directory

The X.500 series of standards specify the methods for providing a global distributed directory service. X.500 directory services are expected to be used in conjunction with X.400 in the future.

Clauses D.1 and D.2 of Annex D of F.435 describe EDI requirements for directory use. In essence, EDI users can benefit from X.500 directory services as follows:

- by obtaining the ORAddress for a given EDI application;

- by obtaining EDI-specific attributes that pertain to a given EDI User Agent (as identified by its ORName).

At present, EDI relations are usually subject to bilateral negotiation and agreement. So long as bilateral agreements are required, use of directory services is not critically required, since partners can exchange the required information (for example, ORAddress) during the bilateral negotiation.

Use of directory services will become a necessity as "open" EDI relations become common. By an "open" EDI relation we mean one in which a company opens its EDI gateway to all potential partners, without negotiating individual bilateral agreements with each partner.

Open trading relations are of course the norm with paper-based business transactions. Any company is willing to accept a Purchase Order from any other company (provided payment is suitably guaranteed).

In the future, it can be expected that companies will be willing to accept EDI Purchase Orders from any *originator*, provided that security services are used to authenticate the *originator*, and provided that payment is suitably guaranteed.

Once open EDI trading relations become common, there will be a significant need for directory services, in order to avoid the manual operations required at present to discover the potential partner's ORAddress, EDI capabilities, etc.

11.1   Obtaining ORAddresses

Existing EDI implementations use a variety of schemes to address a particular EDI application.   The *EDI Name* in carried in the EDI Interchange *header segment*, for example, in the EDIFACT UNB *segment*, or the ANSI X12 ISA *segment*.

Before an EDI Interchange can be transmitted with P_{edi} the *originator* EDI UA must obtain the ORName of the *recipient* UA.

The X.435 standard does not specify how this should be done. Mapping of EDI Names to X.400 ORNames is considered a local matter.

However, X.500 directory services can be used in order to obtain an ORName from an EDI Name.

How this can be done is explained in clauses D.3 and D.4 of Annex D of F.435. Essentially, the aliasing technique of Annex E of X.501 is used.

It is assumed that some organization will maintain a directory sub-tree containing EDI Names (for example, in the USA a commercial organization could maintain a directory of all *DUNS Numbers*. DUNS Numbers are EDI Names issued by the Dun and Bradstreet company. They are commonly used for ANSI X12 EDI Interchanges.)

Given the EDI Name (for example, a DUNS Number), a directory query is issued, and the Directory User Agent will return the Directory Name of the EDI UA corresponding to the EDI Name.

Given the Directory Name of the EDIN UA, a second directory query is issued, and the Directory User Agent will return the ORAddress of the EDI UA corresponding to the Directory Name, and, optionally, the EDI-specific attributes of the EDI UA (see section 11.2).

The following example shows the flow of steps, in the case where a DUNS Number is used as the EDI Name:

1.   The EDI application passes the DUNS Number to the EDI UA. The DUNS Number is "12345".

2.   The EDI UA calls the Directory User Agent with the Directory Name: *Country* = US, *Organization* = DUNS, *EDIUser* = 12345.

3.   The Directory User Agent returns the *aliassed* Directory Name: *Country* = US, *Organization* = TELECOMEDI, *OrganizationUnit* = SALES, *EDIUser* = INVOICES.

4.   The EDI UA calls the Directory User Agent with the
      Directory Name returned in step 3, in order to obtain the
      ORAddress of the *recipient* UA.

EDIUser is a specific *object class* defined in Annex H of X.435. *Object classes* in general are defined in clause 9.4 of X.501.

Figure 11.1 illustrates the process outlined above. A more complete illustration is given in Figure D.2 of Annex D of F.435.

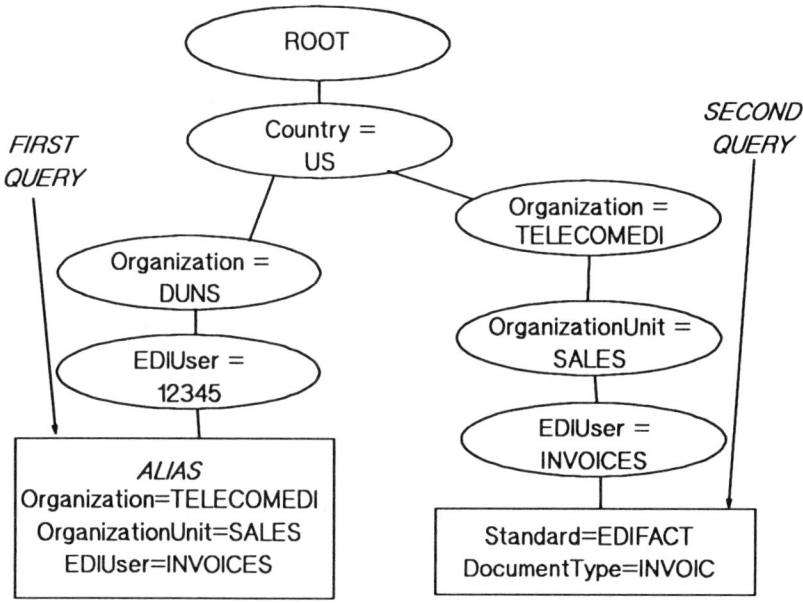

Figure 11.1 - Use of Directory

11.2   Obtaining EDI-specific attributes

As mentioned above, the directory can be used to obtain information on the specific EDI capabilities of a *recipient* EDI UA.

Annex H of X.435 defines the EDI specific attributes which may be held in an X.500 directory. Annex J of X.435 contains descriptive text for the attributes defined in Annex H. The following are worth noting:

EDIBodypartType: EDIFACT, ANSI X12, UN/TDI or other

StandardVersion: Version of the EDI syntax that is supported by the
                     recipient EDI application.

DocumentType: Types of EDI Messages (for example, Invoices) that
                           are supported by the *recipient* EDI application.

DocumentVersion: Versions of the EDI Messages that are supported by
                           the *recipient* EDI application.

Thus, by performing a directory query, an *originator* can discover whether or
not a *recipient* will be able to accept and process the EDI Interchange that
the *originator* intends to send.

12   Areas for bilateral agreements

Due to the diverse needs of the EDI user community, some portions of the P_{edi} protocol specify generic mechanisms that cannot be used in practice without either bilateral agreements between the sender and the receiver, or a functional group or industry usage guideline.

The remaining sections of this Chapter discuss these generic mechanisms.

12.1   Action Request Code

Clause 8.2.3.2 of X.435 defines an *object identifier field* in the EDIM Heading called Action Request Field. Two values are defined for this *field*:

-      For Action

-      Copy

If no value is specified for this *field*, then the value "For Action" is assumed to apply.

The values provided correspond to the well-known concepts of "orginal" and "copy" for paper documents.   For example, an EDIM with Action Request set to "Copy" could be sent to an archiving process.

Additional values may be defined by registering new *object identifiers*.

12.2   EDIN Initiator

Clause 9.1.6 of X.435 defines an integer *field* in the EDIN Common Fields called EDIN Initiator. This *field* can take one of the following values:

-      "internal-UA", which means that the EDIN was generated by the UA itself, without any interaction with the EDI application;

-      "external-UA", which means that the EDIN was generated at the request of the EDI application;

- "internal-MS", which means that the EDIN was generated by the MS itself, without any interaction with the EDI application.

Presence of this *field* is mandatory.

Clause 9.1.6 of X.435 specifies that, irrespective of the value of the EDIN Initiator *field* for a PN EDIN, responsibility for the EDIM is accepted when a PN EDIN is generated.

However, any particular EDI user community, or any two consenting trading partners, could agree to generate only EDINs with EDIN Initiator value "external-UA".

That is, an EDI user community could agree that PN EDINs would only be generated on request of the EDI application, rather than by a UA or MS on its own.

An agreement of this sort would ensure that the EDI application had actually processed at least part of the EDI Interchange contained in the EDIM, before responsibility is accepted.

In effect, an agreement of this sort would define precisely the boundaries that must be crossed before responsibility is accepted, and would make precise the "Fuzzy Border" which is discussed in section 15.4.

12.3   Forwarding responsibility

As we have seen in Chapters 5 and 6, a UA has the option to *forward* responsibility for an EDIM if the Responsibility Passing Allowed *field* of the EDIM Heading has been set to the value "true".

There are circumstances where forwarding of responsibility may be viewed as undesirable, either within a particular EDI user community, or among two trading partners. In this case, nothing prevents the trading partners from agreeing not to permit forwarding of responsibility.

The P_{edi} protocol specified in X.435 is extremely general, since, as explained in Chapter 2, it was designed to meet the very diverse needs of many different types of EDI users.

Individual EDI users and trading communities may well decide to restrict themselves to using a subset of P_{edi}, in order to simplify implementation.

13 Interworking with P2, P1/0 and '84 MTAs

EDIMs and EDINs may be transported by MTAs that conform to the 1984 version of X.400, as explained in section 13.5.

In addition, EDIMs, as defined in X.435, can be converted to and from EDI Interchanges embedded in X.400 IPM Messages using the CEC TEDIS P2 convention, or carried directly within the P1 *envelope* using the ANSI X12/NIST P1/0 conventions.

These conversions may be performed automatically, unless the originator has specified the Implicit Conversion Prohibited service in the P1 *envelope*.

Sections 13.1 to 13.4 present one possible way of performing the conversions. It is to be hoped that EDI user groups, or functional profile groups, will adopt some conversion rules, in order to standardize interworking of future EDI networks based on P_{edi} with existing implementations of X.400 EDI networks not based on the P_{edi} protocol.

13.1 P2 to P_{edi}

It is assumed that there is one and only one *body part* in the IPM Message, and that this *body part* contains an EDIFACT message.

The IPM *body part* becomes the first, and only, *body part* of the EDIM.

The EDIM Heading *fields* are set as follows:

EDIMIdentifier: Originator ORName concatenated with the LocalIPMIdentifier portion of the IPMIdentifier.

Originator: Originator ORName.

Recipients: Recipients from IPM Heading. EDI Notification Requests are not specified (none are requested).

EDIBodyPartType: Value for EDIFACT, ISO 646, or any other valid
                        value if the entity performing the conversion can
                        determine which EDI syntax is contained in the IPM
                        body part and which character encoding is used for
                        the EDI syntax.

Other heading fields may be set if the entity performing the conversion is
capable of parsing the EDI Interchange and discovering the correct values of
the EDIM Heading *fields*.

Since there are no notification requests, the EDI UA will never create an
EDIN when it receives an EDIM converted from P2.

Example 13.1 shows how an EDI Interchange contained in an IPM
Message could be embedded in an EDIM.

Example 13.1 - P2 to/from P_{edi}

IPM Heading	
Field	Value
IPM Identifier	CH/ARCOM/HP/EDI/PURCH, 90/01/26/10:21:04
Originator	CH/ARCOM/HP/EDI/PURCH
Recipient	CH/ARCOM/SUPPLIER
IPM Body	
UNB+UNOA:1+HP+SUPPLIER+90/01/26+135' UNH+1+ORDERS:1' ... UNZ+1+135'	

Example 13.1 - *continued*

EDIM Heading	
Field	Value
This EDIM	CH/ARCOM/HP/EDI/PURCH, 90/01/26/10:21:04
Originator	CH/ARCOM/HP/EDI/PURCH
Recipients Recipient	CH/ARCOM/SUPPLIER
EDI Body Part Type	EDIFACT, ISO 646

EDIM Body
UNB+UNOA:1+HP+SUPPLIER+90/01/26+135' UNH+1+ORDERS:1' ... UNZ+1+135'

13.2 P$_{edi}$ to P2

The first *body part* of the EDIM will be copied to the IPM body. All other *body parts* of the EDIM will be discarded.

The IPM Heading *fields* are set as follows:

IPMIdentifier: EDIMIdentifier.

Originator: Originator ORName.

Recipients: Recipients from EDIM heading. All Recipients become IPM Primary Recipients. An NN EDIN with NN Reason Code set to the value "unspecified" is created for each Recipient where a Notification Request was specified. The EDINOriginator is set to the Recipient ORName. IPM Notifications shall not be requested.

Subject: not present, or set to a single blank character.

If EDINs have been requested, the *originator* will always receive an NN. Since no IPM Notifications are requested, the IPM UA will never create an IPM Notification when it receives an IPM converted from an EDIM.

The example shown in section 13.2 serves also as an example of how an EDI Interchange contained in an EDIM can be embedded in an IPM Message.

13.3   P1/0 to P_{edi}

There is one and only one EDI Interchange contained in the *content* of the X.400 *message*. The *content type field* of the P1 *envelope* has the value "undefined".

The EDI Interchange becomes the first, and only, *body part* of the EDIM.

The EDIM Heading *fields* are set as follows:

EDIMIdentifier: Originator ORName concatenated with the UTCTime at which the conversion from P1/0 to P_{edi} was performed.

Originator: Originator ORName.

Recipients: Recipients from the P1 *envelope*. EDI Notification Requests are not specified (none are requested).

EDIBodyPartType: Value for ANSI X12, EBCDIC if the Encoded Information Type *field* (EIT) of the P1 *envelope* has the value "undefined", value for ANSI X12, ISO 646 if the EIT has the value "IA5String", or any other valid value if the entity performing the conversion can determine which EDI syntax is contained in the *content* and which character encoding is used for the EDI syntax.

Other heading fields may be set if the entity performing the conversion is capable of parsing the EDI Interchange and discovering the correct values of the EDIM Heading *fields*.

Since there are no notification requests, the EDI UA will never create an EDIN when it receives an EDIM converted from P1/0.

Example 13.2 shows how an EDI Interchange contained in a P1 *envelope* could be embedded in an EDIM.

Example 13.2 - P1/0 to/from P$_{edi}$

P1 Envelope	
Field	Value
Originator	CH/ARCOM/HP/EDI/PURCH
Recipient	CH/ARCOM/SUPPLIER
Encoded Information Type	IA5String
Content	
ISA******HP**SUPPLIER*90/01/26***135 ST*850*1 ... IEA*1*135	

Example 13.2 - *continued*

EDIM Heading	
Field	Value
This EDIM	CH/ARCOM/HP/EDI/PURCH, 90/01/26/10:21:04
Originator	CH/ARCOM/HP/EDI/PURCH
Recipients Recipient	CH/ARCOM/SUPPLIER
EDI Body Part Type	ANSI X12, ISO 646
EDIM Body	
ISA******HP**SUPPLIER*90/01/26***135 ST*850*1 ... IEA*1*135	

13.4   P$_{edi}$ to P1/0

The first *body part* of the EDIM will be copied to the *content*. All other *body parts* of the EDIM will be discarded.

The P1 *envelope fields* are set as follows:

ContentType: undefined.

Originator: Originator ORName.

Recipients: Recipients from EDIM heading. An NN EDIN with NN Reason Code set to the value "unspecified" is created for each Recipient where a Notification Request was specified. The EDIN Originator is set to the Recipient ORName.

Encoded Information Type: set to "undefined" if the EDI Body Part Type is encoded with the EBCDIC character set, set to "IA5String" if the EDI Body Part is encoded in ISO 646 (ASCII), not present otherwise.

If EDINs have been requested, the *originator* will always receive an NN.

The example shown in section 13.3 also serves as an example of how an EDI Interchange contained in an EDIM can be embedded in an X.400 *message* using the P1/0 convention.

13.5   Using P$_{edi}$ with 1984 MTAs

Clause 19.2 of X.435 specifies the mechanism by which a P$_{edi}$ UA can *submit* and *receive* EDIMs and EDINs to and from a 1984 version MTA.

NOTE: These provisions override the provisions of Annex B of X.419. A statement to this effect is expected to be added to future versions of the X.400 Implementor's Guide, which is the document that specifies corrections to the published versions of the standards.

The following rules are specified:

-    When an EDIM or EDIN is *submitted* to a 1984 version MTA, the *content type* is set to the integer value "35".

- A P_{edi} UA must be able to accept *content types which are* set either to the integer value "35" or to *the object identifier* value for P_{edi} which is specified in X.435. The *object identifier* is used by 1988 version MTAs, the integer by 1984 version MTAs.

- When a 1988 version MTA *relays* an EDIM or EDIN to a 1984 version MTA, it sets the *content type* to the integer value "35".

In addition, clause 19.4 of X.435 specifies that a P_{edi} UA which *submits* an EDIM to a 1984 version MTA shall set the "undefined" bit of the "basic-encoded-information-type" (called "built-in-encoded-information-types" in the 1988 version), and shall not set the "external-encoded-information-type" *field*.

As a consequence, when EDIMs are submitted to a 1984 version MTA, the final MTA (the one that *delivers* the EDIM) cannot select EDIMs for *delivery* on the basis of the EDI Body Part Type, which contains an indication of the EDI syntax used for the EDI Interchange (for example, ANSI X12, EDIFACT or PRIVATE). This means that the final MTA will *deliver* all EDIMs that were originally *submitted* to a 1984 version MTA, even if the UA attached to the final MTA had requested that EDIMs which contain certain EDI Body Part Types (for example, PRIVATE) not be *delivered* to it.

14 Complications of forwarding

It has been observed that, if an EDIM is forwarded several times, the originator may receive both an NN and a PN for the same EDIM (that is, the EDINs would all contain the same EDIMIdentifier).

Figure 14.1 illustrates one way in which such a situation could arise.

NOTE: The following notation will be used in all figures in this chapter:

EDIMx, where x is a number, denotes the EDIM that was sent by UA1 to UAx.

EDINx, NRNx, PRNx, where x is a number, denotes the EDIN, NN or PN that is sent from a final recipient UA to UA1 for the EDIMx.

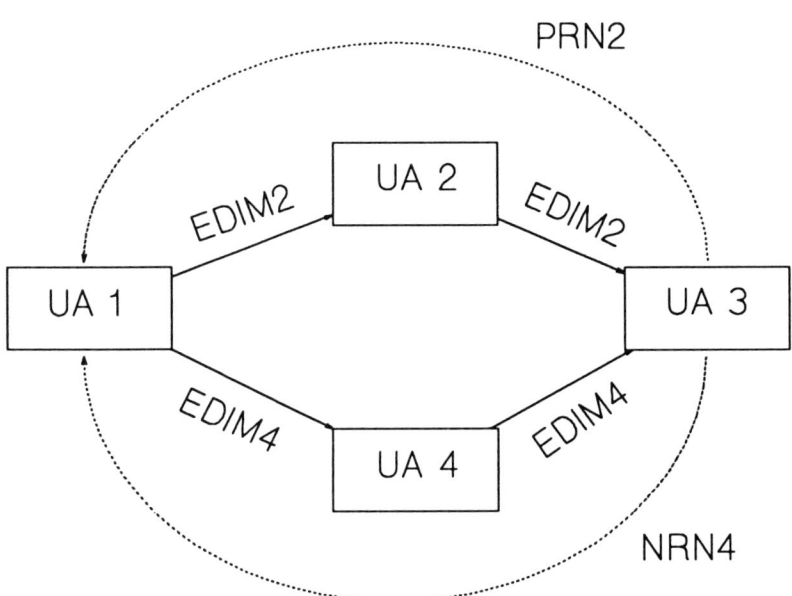

Figure 14.1 - Complications of forwarding 1

UA1 specifies both UA2 and UA4 as recipients, and requests both NN and PN. UA2 does not accept responsibility and forwards the EDIM to UA3; UA3 accepts responsibility and sends a PN to UA1. Later, UA4 also does not accept responsibility and forwards the EDIM to UA3; UA3 refuses responsibility (because it has already received the EDIM) and sends an NN to UA1.

Thus, UA1 receives first a PN, and then an NN from UA3. Both EDINs contain the same EDIMIdentifier, since they both refer to the EDIM created by UA1 and sent by UA1 to UA2 and UA4.

However, UA1 can distinguish the two EDINs because the EDIN contains the ORName of the FirstRecipient. That is, PRN2 in Figure 14.1 contains the ORName of UA2, and NRN4 in the figure contains the ORName of UA4.

Thus, it is perfectly clear to UA1 that the EDIM sent to UA2 was forwarded to UA3, who accepted responsibility, and that the EDIM sent to UA4 was also forwarded to UA3, who refused responsibility because he had already received the same EDIM .

UA1 does need some logic in order to correlate the several EDINs that can be received, however this logic is relatively straight-forward (as explained in section 6.3).

Figure 14.2 illustrates a simpler, but more pathological case. From UA1's point of view, the situation is no more complex than that illustrated in Figure 14.1.

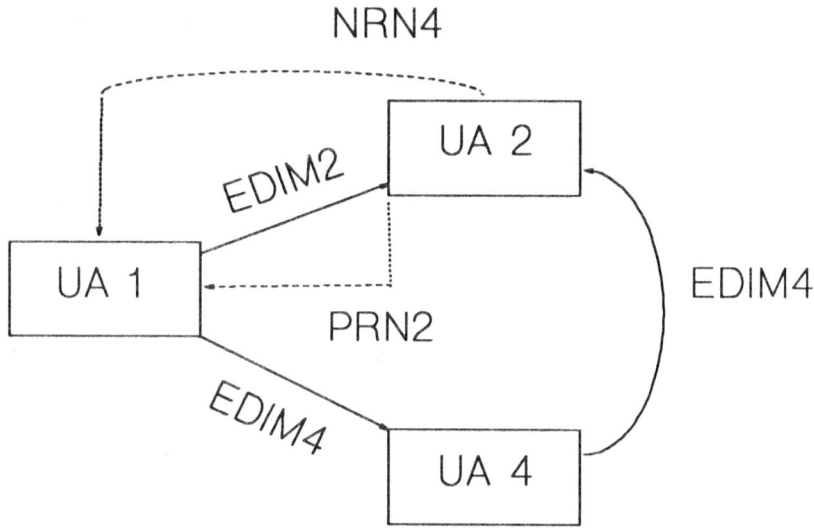

Figure 14.2 - Complications of forwarding 2

UA1 specifies UA2 and UA4 as recipients, and requests notifications. UA2 accepts responsibility and sends a PN. UA4 does not accept responsibility and forwards the EDIM to UA2, who refuses responsibility (because it has already received the EDIM). UA2, having already sent PRN2, also sends NRN4.

Again, UA1 can distinguish the EDINs because they contain the ORName of the FirstRecipient: UA2 for PRN2, UA4 for NRN4.

The situation illustrated in Figure 14.2 can arise in a more complex way, as illustrated in Figure 14.3.

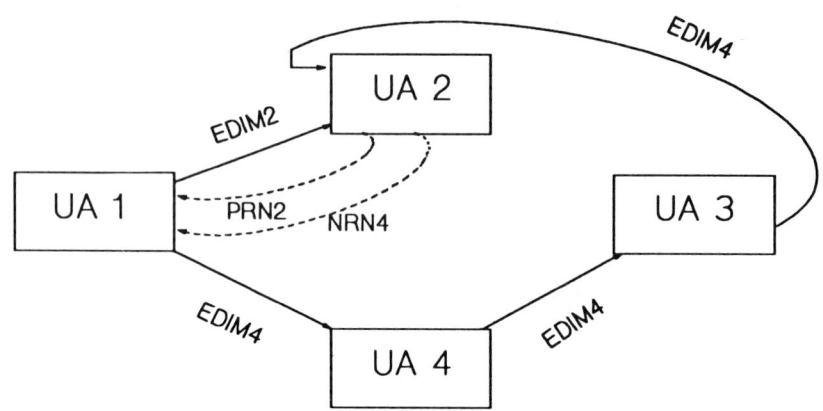

Figure 14.3 - Complications of forwarding 3

In this case UA4 does not accept responsibility and forwards the EDIM to UA3. UA3 also does not accept responsibility and forwards the EDIM to UA2.

From UA1's point of view, there is no difference between the cases illustrated in Figures 14.2 and 14.3.

A more complex variant of the situation illustrated in Figure 14.1 is illustrated in Figure 14.4.

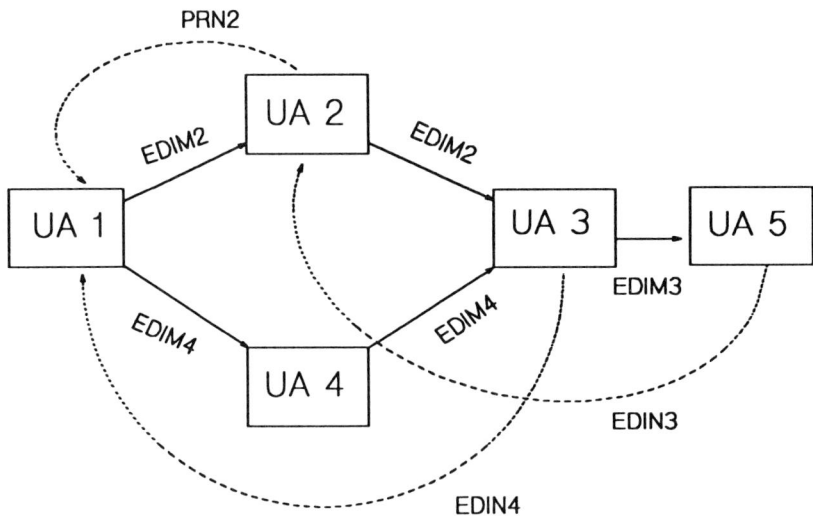

Figure 14.4 - Complications of forwarding 4

In this case, UA1 specifies both UA2 and UA4 as recipients, and requests notifications. UA2 accepts responsibility, and sends a PN to UA1. UA4 does not accept responsibility and forwards the EDIM to UA3. UA3 accepts (or refuses) responsibility and sends an EDIN to UA1.

UA1 receives PRN2 from UA2 and EDIN4 from UA3. Since PRN2 contains the ORName of UA2 and EDIN4 contains the ORName of UA4, UA1 can distinguish the EDINs, and process them correctly.

UA2 forwards the EDIM to UA3.   Since UA2 has accepted responsibility, the EDIMIdentifier of EDIM3 is *not* the same as the EDIMIdentifier of EDIM2.   UA3 does not accept responsibility and forwards EDIM3 to UA5.   UA5 accepts (or refuses) responsibility, and sends EDIN3 to UA2. EDIN3 contains the EDIMIdentifier of EDIM3, and also the ORName of UA3.

Figure 14.5 illustrates a pathological case that is not likely to occur in practice. Again, even if it does occur, the EDINs contain all the information required to allow UA1 to make sense of the EDINs that it receives.

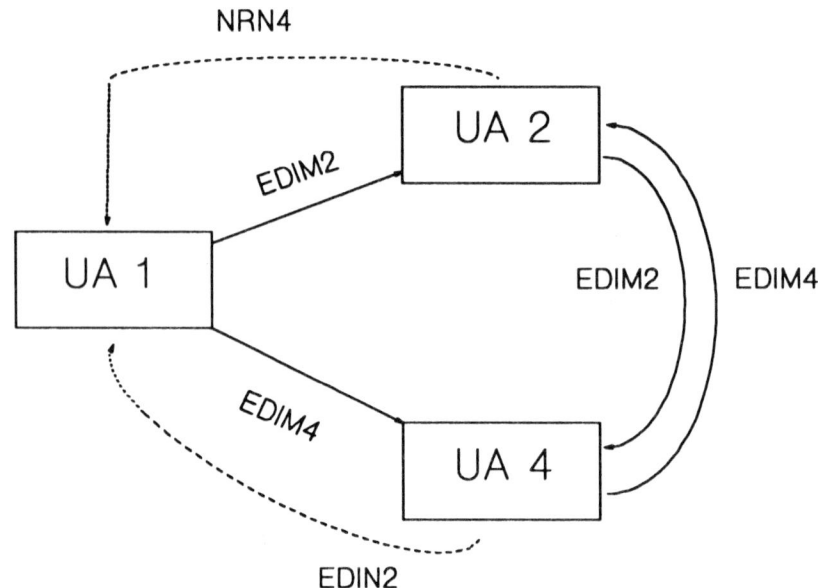

Figure 14.5 - Complications of forwarding 5

UA1 specifies UA2 and UA4 as recipients, and requests notifications.

When UA2 receives EDIM2, it does not accept responsibility and forwards the EDIM to UA4. UA4 accepts (or refuses) responsibility and sends EDIN2 to UA1.

When UA4 receives EDIM4, it does not accept responsibility and forwards the EDIM to UA2. UA2 accepts (or refuses) responsibility and sends EDIN4 to UA1.

Thus, UA1 receives EDIN4 from UA2 and EDIN2 from UA4. That is, the EDIN received from UA2 contains the ORName of UA4 as FirstRecipient, while the EDIN received from UA4 contains the ORName of UA2 as FirstRecipient.

UA1 thus knows that the EDIN received from UA2 does <u>not</u> refer to the EDIM that UA1 sent to UA2, but rather to the EDIM that UA1 sent to UA4, and similarly the EDIN received from UA4 refers to the EDIM sent to UA2.

Figure 14.6 illustrates how the pathological case of Figure 14.5 might actually arise in practice.

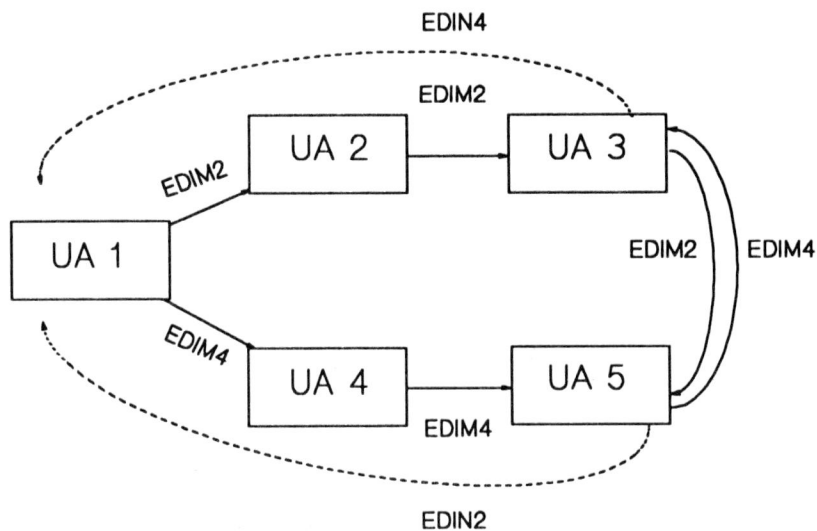

Figure 14.6 - Complications of forwarding 6

The situation is the same as that of Figure 14.5, except that an extra stage of forwarding is introduced.

UA1 receives the EDIN for EDIM2 from UA5, and the EDIN for EDIM4 from UA3. Since each EDIN contains the ORName of the FirstRecipient, UA1 has no difficulty in correlating the EDINs actually received with the EDINs requested and pending.

To conclude, it can be noted that UA1 will have no difficulty in correlating the received EDINs with the expected EDINs if it keeps track of the *recipients* from which notifications have been requested, as suggested in section 6.3.

15   Other topics

15.1   Hiding Distribution Lists

The P_{edi} protocol does not allow the Blind Copy (BCC) feature of IPM. In IPM, selected *recipients* can be designated as BCC; the list of these *recipients* is not disclosed to the other *recipients* of the IPM.

It is recognized that hiding recipients is a desirable feature, for example, when sending a Request for Quotation.

However, it was felt that the BCC feature of IPM did not provide the desired functionality.

Instead, the EDI UA should *submit* the EDIM multiple times, if certain *recipients* should be hidden from others. Each time the EDIM is *submitted*, it should contain only those *recipients* who are allowed to know each other.

15.2   Use of Per-Recipient Fields

Many of the *data elements* present in the EDI Header *segments* (for example, the EDIFACT UNB *segment*) can be specified on a per-*recipient* basis in P_{edi}, even though these *data elements* appear once and only once in the EDI Header *segment*.

Clause 8.2.3 of X.435 defines the per-*recipient fields* of the EDIM Heading.

The rationale behind allowing different values of these *fields* for each *recipient* is to allow P_{edi} to cater for possible future extensions of the EDI Header *segments*.

It is thought likely that the EDI Header *segments* will be expanded to allow specification of multiple EDI Receiver *data elements*. If and when this expansion takes place, no changes will be required to X.435, since X.435 already allows specification of EDI Header *segment data elements* on a per-*recipient* basis.

Even in the absence of changes to existing EDI syntaxes, a sophisticated EDI UA could generate multiple *recipients* for an EDI Interchange such as Request for Quotation that would naturally be sent to several *recipients*.

Of course, until the EDI syntaxes are expanded, the *data elements* contained in the EDI Header *segment* of the EDI Interchange would be consistent with the per-*recipient fields* for one and only one *recipient*, and would be inconsistent with the per-*recipient fields* for other *recipients*.

In this case, the receiving EDI application would have to ignore the *data elements* contained in the EDI Header *segment* of the EDI Interchange, and rely on the values contained in the EDIM Heading instead.

Example 15.1 illustrates the situation:

Example 15.1 - Use of Per-Recipient Fields

EDIM Heading	
Field	Value
This EDIM	CH/ARCOM/HP/EDI/CATLG, 90/01/26/10:21:04
Originator	CH/ARCOM/HP/EDI/CATLG

Recipients
  Recipient
  Interchange Recipient
  Interchange Control Reference
	
	CH/ARCOM/CUSTOMER1
	CUSTOMER1
	14

Recipient
  Interchange Recipient
  Interchange Control Reference
	CH/ARCOM/CUSTOMER2
	CUSTOMER2
	76

EDI Body Part Type	EDIFACT, ISO 646
EDI Message Type	PRICAT
Interchange Sender	HP
EDIM Body	
UNB+UNOA:1+HP+*+90/01/26+0' UNH+1+PRICAT:1' ... UNZ+1+0'	

NOTE: In the UNB *header segment* of the EDI Interchange, the Interchange Recipient is set to the value "*". This indicates that the actual Interchange Recipient is to be found in the Interchange Recipient *field* of the EDIM Heading (values "CUSTOMER1" and "CUSTOMER2").

Similarly, the Interchange Control Reference is set to the value "0" in the UNB *header segment* of the EDI Interchange, to indicate that the actual Interchange Control Reference is found in the Interchange Control Reference *field* of the EDIM Heading (value "14" for CUSTOMER1, value "76" for CUSTOMER2).

15.3   Selective Redirection

Redirection is defined as the action the MTA takes when it systematically sends all *messages* addressed to a particular ORAddress to some other ORAddress.

Selective Redirection is defined as *redirection* of only some messages, which are selected for *redirection* on the basis of *fields* in the EDIM Heading.

It is clear from the discussion in Chapters 5 and 7 that *Selective Redirection* can be implemented by a sophisticated EDI User Agent.

An MTA normally accesses only the P1 *envelope*, and thus cannot perform *Selective Redirection*.

However, it should be noted that existing X.400 standards do not specify the internal working of an MTA, and that the standards do not therefore forbid an MTA performing *Selective Redirection* based on data contained in the EDIM Heading, provided that such *Selective Redirection* is viewed as a local implementation; the method for informing the MTA of the *Selective Redirection* criteria is not standardized, and is not part of the X.400 standards.

That is, an EDI user would use a proprietary interface to instruct the MTA on the *Selective Redirection* criteria, if the MTA can provide such a function.

No changes to the P1 protocol or any other X.400 standards are required.

15.4   The fuzzy border

An exact definition of the boundary between the UA and the EDI process is beyond the scope of CCITT Recommendations.

Thus, the boundary between the UA and the EDI process is a matter of local implementation. The EDI application can include a process that is completely unrelated to OSI standards. For example, the EDI process could perform a conversion from X.400 to a proprietary telecommunications protocol.

The concept of *EDI Responsibility* was discussed in Chapter 6. *EDI Responsibility* is accepted when an EDIM leaves the X.400 world. Since the boundary between the UA and the EDI process is not defined in any CCITT Recommendation, it is a matter of local implementation to determine where in the process of moving an EDIM from the UA to the EDI process the *EDI Responsibility* is accepted.

For example, if the EDI process performs a conversion from X.400 to a proprietary telecommunications protocol, *EDI Responsibility* could be accepted either at the beginning of the conversion, or only after some confirmation that the transmission using the proprietary protocol has effectively taken place.

It should be noted that the fuzzy border between the UA and the EDI process is not well represented by a line. It is better thought of as something with substantial thickness, as illustrated in Figure 15.1.

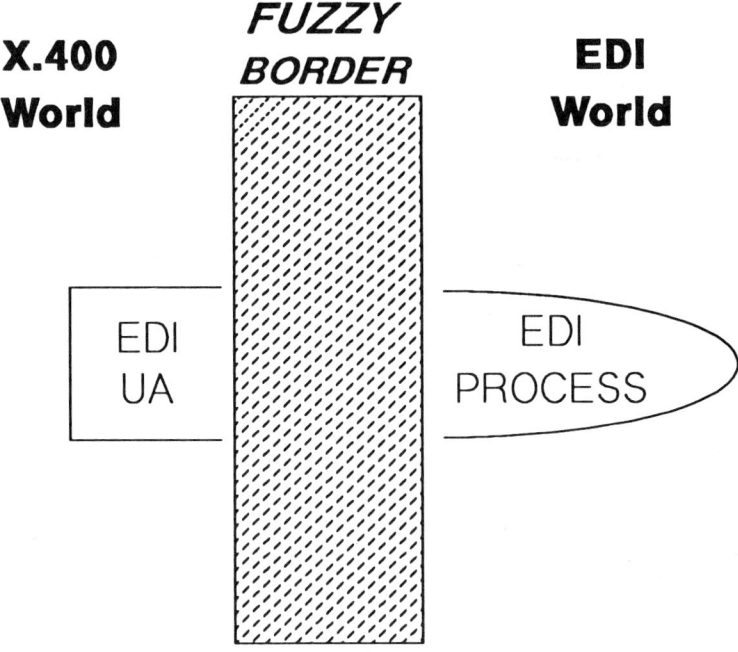

Figure 15.1 - The fuzzy border

The thickness of the fuzzy border is due to the fact that substantial processing can take place within this border. For example, the processing required to confirm transmission using a proprietary protocol can take place within the fuzzy border.

Clause 9.1.6 of X.435 defines the EDIN Initiator. This *field* permits the *receiving* UA to send to the *originating* UA information regarding which side of the fuzzy border the EDIN came from. The EDIN Initiator can be specified for PN, NN or FN EDINs.

If the EDIN Initiator has the value "external-UA", then the EDIN was generated on the EDI application side of the fuzzy border.

If the EDIN Initiator has the value "internal-UA", then the EDIN was generated on the UA (X.400) side of the fuzzy border.

Clause 9.1.6 of X.435 specifies that, irrespective of the setting of the EDIN Initiator *field, EDI Responsibility* is deemed to be accepted whenever a PN EDIN is generated.

See section 12.3 for a discussion on how the EDIN Initiator *field* can be used in practice.

15.5   P$_{edi}$ API

Since EDI is, by definition, direct computer to computer exchange of data, a P$_{edi}$ User Agent will normally interface with an EDI application through some sort of programmatic interface. That is, the EDI application would call a programmatic interface in order to pass data to, and receive data from, the EDI UA.

Such programmatic interfaces are called Application Program Interfaces (APIs).

Clause 12 of X.435 defines ASN.1 constructs called *Abstract Operations*. These constructs do not correspond to any data that are actually transferred with P$_{edi}$; rather, the *Abstract Operations* provide a suggestion for defining an API.

Definition of an API is beyond the scope of CCITT Recommendations, and will likely be undertaken by a functional profile group, or by an industry group such as the API Association. We show here some primitive functions that could form part of an API for P$_{edi}$ (the functions must of course comply with all the rules of X.435, which are outlined in the previous chapters of this book; for example, they must comply with the security provisions outlined in Chapter 10, if security is requested):

Originate-EDIM: This function would have the following arguments:

- values for the EDIM Heading *fields*;

- one or more *body parts*, of which the first would have to contain an EDI Interchange.

Read-EDIM-Heading: This function would have as argument an EDIMIdentifier.

It would return to the EDI application all of the values of the EDIM Heading for the EDIM in question.

The EDI application could use the returned values in order to decide what subsequent actions to take (for example, Receive-EDIM or Originate-EDIN of type NN).

NOTE: Provision of this primitive function provides the EDI application with the possibility of making decisions on the basis of the contents of the EDI Heading *segment*, without having to accept *Responsibility* for the EDIM, and without having to parse the EDI Interchange itself.

Receive-EDIM: This function would have as argument an EDIM Identifier.

It would return the EDIM, composed of a Heading and one or more *body parts*.

Originate-EDIN: This function would have the following arguments:

- indication of whether to *originate* a PN, NN or FN;

- values for the EDIN Common *fields*. Note that EDIN Initiator would be set to the value "external" automatically, since, by definition, the EDIN is being generated outside the X.400 world;

- value for the Reason Code, if required.

Receive-EDIN: This function would have as arguments a Subject EDIM (the EDIM Identifier of the EDIM to which the EDIN refers - see clause 9.1.1 of X.435) and a First Recipient.

It would return an indication of the type of the EDIN received (PN, NN or FN) and the *fields* contained in the EDIN.

NOTE: Provision of all four of the basic functions Originate/Receive EDIM/EDIN allows creation of processes that could fail to conform with the requirements of X.435. For example, an EDI application could invoke Receive-EDIM and then fail to invoke Originate-EDIN, even though an NN or PN Notification was requested, and must be generated in order to conform to X.435.

Forward-Body-Parts: This function would have the following arguments:

- a list of the numbers of the *body parts* to be forwarded;

- additional *body parts* to be added;

- values for the EDIM Heading *fields* (for example, new *recipients*).

Body parts not contained in the list of *body parts* to be forwarded would be dropped.

A PN EDIN <u>must</u> be generated if this function is invoked and PN Notification was requested.

Forward-Unchanged: This function would have the following argument:

- values for the new *recipients*.

An FN EDIN <u>must</u> be generated if this function is invoked and FN Notification was requested.

15.6   Physical Delivery

Clauses 15.4 of X.435 and 11 of F.435 define a Physical Delivery Access Unit (PDAU). This is an Access Unit that is capable of printing an EDIM.

That is, a service provider could use this facility to print an EDI Interchange contained in an EDIM, and send the printed copy to the final recipient, if the recipient is unable to receive X.400 *messages*.

A PDAU cannot send a PN EDIN if EDINs are requested. It can send an NN EDIN if it cannot perform the physical rendition (printing). It must send an FN EDIN if it can perform the physical rendition (printing). X.435 specifies NN and PN reason codes to cover these cases (see clauses 9.3.1 and 9.4.2 of X.435).

15.7   Conformance

Clause 21 of X.435 specifies the requirements that an implementation must meet if it claims to conform to the standard. The language in this clause is rather generic, and there are no specific lists of features that must be implemented.

More specific guidance for implementors will be contained in forthcoming PICS pro-forma statements and tables, which are expected to include specific lists.

It is expected that conformant applications will be required to provide support for all of the P_{edi} protocol elements that correspond to the following Elements of Service defined in F.435:

- all Elements of Service in Table 5 of F.435 (clause 14.1 of F.435);

- Elements of Service classified as Essential (E) in Table 6 of F.435 (clause 14.2 of F.435).

The following points are worth noting:

- Physical Delivery is not an Essential Element of Service, so its implementation will not be required for conformance;

- EDI Forwarding is not an Essential Element of Service, so its implementation will not be required for conformance;

- Security-related Elements of Service are not Essential, so their implementation will not be required for conformance;

- Several Elements of Service (in particular, those related to *forwarding* and cross-referencing) are Essential on *reception*, but not Essential on *origination*.

    Since the related protocol elements must be implemented on *reception* by conformant MTAs and UAs, there should be no interworking problems in the future: any P_{edi} implementation will be able to receive any EDIM.

    However, since the related protocol elements are optional on *origination*, implementors will be able to provide simplified versions of a P_{edi} UA, which will be conformant, even though it lacks the ability to implement certain features.

16 Road map to F.435 and X.435

This chapter contains a road map to the CCITT Recommendations F.435 and X.435, which, taken together, define the P_{edi} protocol. The chapter is organized in terms of major topics, and lists the portions of F.435 and X.435 which should be read in order to understand how P_{edi} deals with the topic.

16.1 Introduction

F.435: clauses 6 and 7 introduce the P_{edi} protocol, explain its relation to other X.400 protocols, and show the structure and relation of key concepts. Clauses 3, 4 and 5 contain definitions. Annex A contains a glossary.

X.435: annexes K and L cover EDI terminology, its relation to the terminology used in X.435 and the relation of the terms used in F.435 with those used in X.435. Clauses 3, 4 and 5 contain definitions.

16.2 Structure of EDIM

X.435: clauses 7.1 and 8 define an EDIM. Per-recipient *fields* are defined in clause 8.2.3.

16.3 Structure of EDIN

X.435: clause 9 defines an EDIN.

16.4 Forwarding, Responsibility and Notifications

F.435: clause 8 defines *responsibility* and *forwarding* and shows the flow of *notifications* when EDIMs are *forwarded*.

X.435: clause 17.3.3 covers the rules related to construction of an
        EDIM when *forwarding* (see in particular 17.3.3.1, 17.3.3.2
        and 17.3.3.4). Clauses 17.3.1.1, 17.3.2.1 and 17.3.3.6 cover the
        rules related to construction of EDINs (respectively PN, NN
        and FN).

16.5   Cross-Referencing

F.435: annex E contains a tutorial on cross-referencing.

X.435: clause 8.2.12 defines the structure of the Cross Referencing
        Information *field*.

16.6   Message Store

X.435: clause 18 defines Message Store *attributes* and *auto-actions*.

16.7   Security

F.435: clause 10 and annex C provide an overview of the security
        functions of P_{edi}.

X.435: clauses 8.2.3.3 and 8.2.12 define the security-related *fields* of
        the EDIM. Clause 9.1.5 defines the security-related *fields* of
        the EDIN. Clause 17.1.3 defines the rules for using these
        fields. Annex I specifies enhancements to the security model
        defined in X.402.

16.8   Use of X.500 Directory

F.435: clause 9 and annex D provide an overview of the EDI use of
        Directory functions.

X.435: annexes H and J define P_{edi}-specific Directory *object classes*
        and *attributes*. Annex M specifies how Directory entries can
        be realized.

16.9   Application Program Interface

X.435:   clause 12 defines *abstract operations*, which can be taken as a starting point for definition of an API.

16.10   Conformance

X.435:   clause 21 specifies conformance requirements.

16.11   ASN.1 definitions

X.435:   annexes A, B, C, D, E, F, G and H contain the reference ASN.1 definitions of the protocol. ASN.1 definitions given in the text of X.435 should be considered explanatory.

16.12   List of Clauses of X.435

The published table of contents for X.435 does not contain the list of all the clauses. The following complete list may prove useful as an addendum to X.435.

16.13 Alphabetic list of X.435 clauses defining the EDIM and EDIN fields

Clauses 7, 8 and 9 of X.435 define the structure and *fields* of the EDIM and EDIN. In almost all cases the name of the *field* is the same as the clause title. Therefore, the following list of clauses, sorted by clause title, can be useful for cross-referencing and look up.

8.2.3.9	Acknowledgement Request
8.2.3.2	Action Request
8.2.18	Application Reference
8.2.3.12	Authorization Information
8.3	Body Part Types
7.	Common data types
9.1	Common Fields
8.2.3.10	Communications Agreement Id
8.2.12	Cross Referencing Information
8.2.17	Date and Time of Preparation
8.2.11	EDI Application Security Elements
8.3.1	EDI Body Part
8.2.6	EDI Body Part Type
8.2.13	EDI Message Type
8.	EDI Messages
9.1.2	EDI Notification Originator
8.2.3.3	EDI Notification Requests
9.	EDI Notifications
8.3.2	EDIM Body Part
7.1	EDIM Identifier
9.1.6	EDIN Initiator
8.2.4	EDIN Receiver
8.2.8	Expiry Time
7.2	Extensions
8.3.3	Externally Defined Body Parts
9.1.3	First Recipient
8.2.5	Forwarded Indication
9.4.4	Forwarded Notification Extensions
9.4	Forwarded Notifications
9.4.2	Forwarded Reason Notification
9.4.1	Forwarded To
9.4.3	FN Supplementary Information
8.2.19	Heading Extensions
8.1	Heading Field Component Types
8.2	Heading Fields
8.1.1.1	Identification Code
8.1.1.2	Identification Code Qualifier

16.14 Alphabetic list of X.435 clauses defining UA operation

Clause 17 of X.435 defines rules related to the construction of EDIMs and EDINs, including rules related to *forwarding*. The following list of clauses, sorted by clause title, can be useful for cross-referencing and look up.

Postscript

The group that defined the P_{edi} protocol, the CCITT Associate Rapporteur Group for EDI and X.400, chaired by Ted Myer, was composed of a combination of X.400 experts, EDI experts and EDI practitioners.

During the early stages of the work, it became apparent that individual members of the group were using significantly different conceptual models when approaching design questions.

X.400 experts tended to view the world as illustrated in Figure P.1.

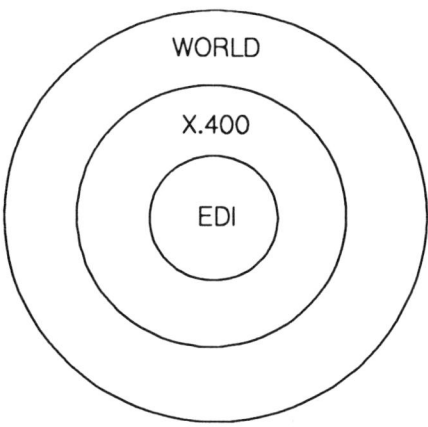

Figure P.1 - X.400 view of the world

That is, EDI was considered to be one type of data to be transmitted within X.400. This view tended towards imposing X.400 conventions on EDI users.

EDI practitioners tended to view the world as illustrated in Figure P.2.

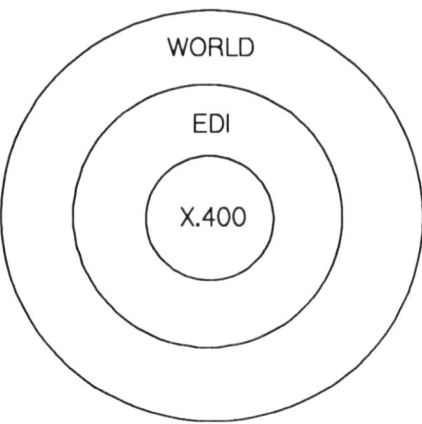

Figure P.2 - EDI view of the world

That is, X.400 was considered to be one type of transport mechanism for EDI data. This view tended towards imposing EDI conventions on X.400 designers.

Taken in their extreme form, the two views are incompatible, and can lead to serious conflicts when designing a protocol.

However, it must be recognized that both views are valid, as illustrated in Figure P.3.

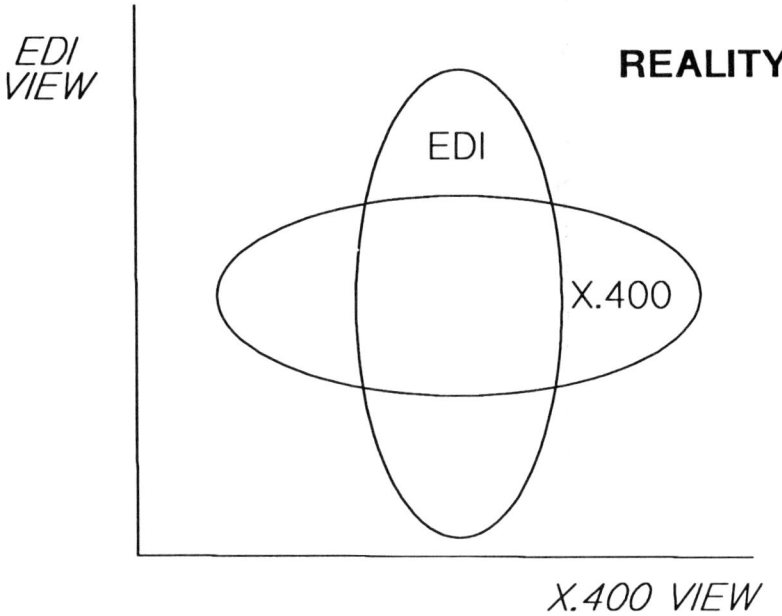

Figure P.3 - Cosmological model

As the caption indicates, this model was colloquially known as the "cosmological model" during the work of the Associate Rapporteur Group. It turned out to be expedient to agree that the general conceptual model of Figure P.3, which embodies both views of the world, could be a useful guide when designing P_{edi}.

When looking at things from the X.400 point of view, EDI appears to be a subset of X.400. When looking at things from the EDI point of view, X.400 appears to be a subset of EDI.

In reality, as shown by the cosmological model, both X.400 and EDI are partly overlapping subsets of the real world, and neither is a subset of the other.

Recognition of the fact that EDI is not a subset of X.400 lead to the fundamental decision to accept for transmission in X.400 EDI Interchanges encoded in character-based EDI syntaxes (like EDIFACT), instead of insisting that all EDI Interchanges to be transmitted in X.400 be encoded in ASN.1.

Glossary

This glossary contains the definitions of the *italicized terms* used in this book. It does not contain the normal English definition of words such as forward, responsibility, user.

See also the glossary contained in Annex A of F.435.

abstract operation	A definition of operations that can be used to specify an interface between applications and X.400 protocols. See section 15.5.
ADMD	A set of MTAs that are owned and operated by a public or private organization on behalf of others.
alias	X.500 Directory technique for indirect lookup of entries. See Annex E of X.501.
ANSI X12	US standard for the structuring of EDI data.
API	Application Program Interface. A standardized set of procedure calls that can be used to interface applications with telecommunications protocols.
API Association	A consortium of corporations that develop API specifications for X.400 protocols.
application	A set of computer programs that serves a specific purpose (for example, order processing).
ASCII	Commonly used character set standard.
ASN.1	Formal language for specifying data structures. Used in X.400, X.500 and other telecommunications standards.
attribute	In Directory see Chapter 11.
attribute	In Message Store see Chapter 9.

auto-action	In Message Store see Chapter 9.
auto-forward	In Message Store see Chapter 9.
body	A set of *body parts* that follows the *heading,* and that, taken together with the *heading,* comprises a *content.*
body part	A structured set of *fields* containing the information that users wish to transmit via X.400.
CCITT	International organization that creates and publishes telecommunications standards, including X.400: International Telephone and Telegraph Consultative Committee.
clearing house	An organization that provides collection, routing and distribution services on behalf of other organizations.
content	A structured set of *fields* that follows the *envelope,* and that, together with the *envelope,* comprises an X.400 *message.* A *content* consists of a *heading* and a *body.*
content type	Specification of the structure applicable to the *content* contained in an X.400 *message* (for example, IPM, EDIM).
data element (EDI)	A set of bits that hold data of a certain type.
deliver	Moving an X.400 *message* from the MTA to the UA.
Directory	See X.500.
Directory Name	A hierarchical name, as defined in the X.500 series of Recommendations. Similar to an ORName.
Directory User Agent	UA that implements the X.500 directory protocols.
Dun and Bradstreet	Company which issues DUNS Numbers.
DUNS Number	An EDI Name issued by the Dun and Bradstreet company.
EBCDIC	Character set defined by the IBM company. Commonly used for ANSI X12.

DL →

DL-expansion : Distribution List expansion performed by an MTA, which generates several recipient ORN from single ORName contained in the P1 envelope

EDI	Electronic Data Interchange. Computer-to-computer exchange of data related to commercial transactions using agreed upon formats and networks.
EDI Name	An alpha-numeric string which identifies a particular EDI application.
EDI Interchange	See *interchange*.
EDI UA	UA that implements the P_{edi} protocol.
EDI User	Specific term for an *object class* defined in X.435. See Chapter 11. Do not confuse with generic term *user*.
EDIFACT	International standard for the structure of EDI data (ISO 9735).
EDIM	An X.400 *message* that contains an EDI Interchange.
EDIMS	EDI Messaging System: P_{edi}.
EDIN	An X.400 *message* sent to the original *originator* to inform him of the disposition of his EDIM (*forwarded*, *responsibility* accepted or *responsibility* refused). See also FN EDIN, PN EDIN and NN EDIN.
element of service	Particular features, functions or capabilities of X.400.
envelope	A structured set of *fields* that comprises the beginning of an X.400 *message*, and that contains the information required by the MTA. The *envelope* is followed by a *content*.
field	A set of bits that hold a certain data type.
FN EDIN	EDIN informing the *originator* that his EDIM has been *forwarded*.
forward EDIM	Action performed by the UA when it packages the received EDIM in a new *envelope*, and *submits* it to the MTA for *relaying* to another UA. See sections 3.4 and 3.5.

forward responsibility Action performed by the UA when it sets protocol elements to inform the new *recipient* UA that it must generate any *notifications* requested by the original *originator*. See section 5.4.

functional group   A set of EDI messages that are grouped together. An *interchange* may contain several *functional groups*.

F.435   CCITT Recommendation that specifies the EDI Messaging Service, that is, the service requirements that P_{edi} is expected to meet.

hash   A computation that reduces a large number of bits to a smaller number of bits, in such a way that all of the original bits influence the outcome of the computation. See Annex D of X.509.

header segment   The first *segment* in an EDI Interchange, typically containing the EDI sender and receiver *data elements*.

heading (X.400)   A structured set of *fields* that comprises the beginning of the *content* of an X.400 *message*, and that contains the information required by the UA. The *heading* is followed by a *body*.

IA5String   An ASN.1 type that corresponds roughly to an ASCII character string.

import   Formal technique used in X.400 standards to define an ASN.1 construct in terms of constructs defined elsewhere. For example, in X.435 *ORName* is *imported* from X.411; that is, the definition of *ORName* is to be found in X.411, not in X.435.

interchange (EDI)   Largest standardized aggregate of EDI data. An *interchange* contains several EDI *messages*.

IPM   An X.400 *message* that contains an inter-personal message (memo).

IPM UA   UA that implements the P2 protocol.

ISO   International organization that creates and publishes standards: International Standards Organization.

ISO 646   ISO standard specifying an ASCII character set used in EDIFACT.

message (EDI)   A set of *segments* that, taken together, comprise a single commercial transaction (for example, purchase order).

message (X.400)   A structured set of data that is *relayed* by MTAs, and *delivered* to a UA. An X.400 *message* contains an *envelope* and a *content*.

message store   See Chapter 9.

MHS   Message Handling System. See X.400.

MOTIS   ISO term for MHS. The ISO MOTIS Standards are essentially identical to the CCITT MHS Recommendations.

MS   Message Store. See Chapter 9.

MS Attribute   See Chapter 9.

MTA   Software that moves X.400 *messages* from one UA or MTA to another UA or MTA.

NIST   US standards-making body: National Institute for Standards and Technology (formerly NBS: National Bureau of Standards).

NN EDIN   EDIN informing the *originator* that *responsibility* for his EDIM has been refused.

notification   X.400 *messages* that are used to carry information related to the disposition of other X.400 *messages*. See also EDIN.

PICS (in conformance) See section 15.7

object class   X.500 construct used to define directory entries. See clause 9.4 of X.501.

object identifier   X.400 construct used to identify certain data types.

octet   Byte.

octet string   A binary string.

originate   The action of creating an X.400 *message*.

originator   UA that sends an X.400 *message*.

ORAddress   X.400 address (includes routing information).

ORName   X.400 address.

OSI   Model for specifying international telecommunica-
                      tions standards: Open Systems Interconnection. The
                      model defines seven layers of operation. The X.400
                      Recommendations fall into the upper layers of the
                      OSI model.

P1   Protocol that specifies how MTAs communicate, and
                      that specifies the X.400 *envelope*.

P1/0   NIST (US) convention for sending EDI messages
                      with the 1984 version of X.400.

P2   Protocol designed for structuring and transmitting
                      inter-personal messages (memos). Also, TEDIS
                      convention for sending EDI messages with the 1984
                      version of X.400.

PDAU   Physical delivery access unit. See section 15.6.

P$_{edi}$ Protocol designed for structuring and transmitting
                      EDI   Interchanges.   Specified   in   CCITT
                      Recommendations F.435 and X.435.

PN EDIN   EDIN informing the *originator* that *responsibility* for
                      his EDIM has been accepted.

PRMD   A set of MTAs that are owned and operated by a
                      private organization on its own behalf.

PTT   Public provider of telecommunication services.

receive   The action of receiving an X.400 *message*. Normally,
                      this implies that the *contents* of the *message* move
                      outside of the X.400 world.

recipient	UA to whom an X.400 *message* is sent.
redirect	Causing an MTA to send all X.400 *messages* intended for one UA to some other UA.
register (in MS)	See section 9.4.
relay	Moving an X.400 *message* from one MTA to another MTA.
remove	To omit a *body part* when *forwarding* a *message*. See section 3.5
responsibility	See the beginning of Chapter 6.
segment (EDI)	A structured set of related *data elements* (for example, a line item in a purchase order).
submit	Moving an X.400 *message* from the UA to the MTA.
TEDIS	Program to promote EDI within the European Economic Community. Funded and managed by Directorate General XIII of the Commission of the European Community.
TRADACOMS	Version of the UN/TDI syntax for EDI Interchanges. Commonly used in the UK.
transaction set	ANSI X12 term for EDI *message*.
UA	Software that moves X.400 *messages* from users to the MTA.
UN/TDI	United Nations standard for the structuring of EDI data. Widely used in the UK in the TRADACOMS version.
UNB	*Interchange* header *segment* in EDIFACT.
user (P_{edi})	EDI process or application that interfaces with an EDI UA.
UTCTime	ASN.1 construct which specifies universal time (GMT).

VAN   Value Added Network.   Provides high-level,
                           specialized services in addition to data transmission.
                           See *clearing house*.

X.25   CCITT Recommendations for transmitting packets
                           of bytes.   Commonly used for the lower-level
                           network protocols in implementations of X.400.

X.400   CCITT Recommendations for structuring and
                           transmitting electronic mail messages.

X.400 message   See *message*.

X.435   CCITT Recommendation that specifies the EDI
                           Messaging System, that is, the P_{edi} protocol that
                           meets the service requirements of F.435.

X.500   CCITT Recommendations for a global distributed
                           directory service.

References

The following documents and books provide useful references (CCITT refers to the International Telephone and Telegraph Consultative Committee, in Geneva, Switzerland; ISO refers to the International Standards Organization in Geneva, Switzerland).

ISO standards can be obtained from local national standardization bodies (for example, AFNOR in France, BSI in the U.K., DIN in Germany).

CCITT Recommendations can be obtained directly from the CCITT Secretariat, Place des Nations, CH-1211 Geneva, Switzerland. F.435 and X.435 should be available from CCITT in mid-1991.

- [CCITT Recommendation X.208 | ISO 8824], Specification of Abstract Syntax Notation One (ASN.1), 1988.

- [CCITT Recommendation X.209 | ISO 8825], Specification of Basic Encoding Rules for Abstract Syntax Notation One (ASN.1), 1988.

- [CCITT Recommendation X.400 | ISO 10021-1], [Recommendation for Message Handling (MHS) | International Standard for Information Processing Systems - Text Communication - Message-Oriented Text Interchange Systems (MOTIS)]: Systems and Service Overview, 1988.

- [CCITT Recommendation X.402 | ISO 10021-2], [Recommendation for MHS | International Standard for MOTIS]: Overall Architecture, 1988.

- CCITT Recommendation X.403, Recommendation for MHS: Conformance Testing, 1988.

- [CCITT Recommendation X.407 | ISO 10021-3], [Recommendation for MHS | International Standard for MOTIS]: Abstract Service Definition Conventions, 1988.

- CCITT Recommendation X.408, Recommendation for MHS: Encoded Information Type Conversion Rules, 1988.

- [CCITT Recommendation X.411|ISO 10021-4], [Recommendation for MHS|International Standard for MOTIS]: Message Transfer System: Abstract Service Definition and Procedures, 1988.

- [CCITT Recommendation X.413|ISO 10021-5], [Recommendation for MHS|International Standard for MOTIS]: Message Store: Abstract Service Definition, 1988.

- [CCITT Recommendation X.419|ISO 10021-6], [Recommendation for MHS|International Standard for MOTIS]: Protocol Specifications, 1988.

- [CCITT Recommendation X.420|ISO 10021-7], [Recommendation for MHS|International Standard for MOTIS]: Inter Personal Messaging, 1988.

- [CCITT Recommendation X.500|ISO 9594-1], [Recommendation for Directory|International Standard for Information Processing Systems - Open Systems Interconnection - The Directory (Directory)]: Overview of Concepts, Models, and Service, 1988.

- [CCITT Recommendation X.501|ISO 9594-2], [Recommendation for Directory|International Standard for Directory]: Models, 1988.

- [CCITT Recommendation X.511|ISO 9594-3], [Recommendation for Directory|International Standard for Directory]: Abstract Service Definition, 1988.

- [CCITT Recommendation X.518|ISO 9594-4], [Recommendation for Directory|International Standard for Directory]: Procedures for Distributed Operation, 1988.

- [CCITT Recommendation X.519|ISO 9594-5], [Recommendation for Directory|International Standard for Directory]: Protocol Specifications, 1988.

- [CCITT Recommendation X.520|ISO 9594-6], [Recommendation for Directory|International Standard for Directory]: Selected Attribute Types, 1988.

- [CCITT Recommendation X.521|ISO 9594-7], [Recommendation for Directory|International Standard for Directory]: Selected Object Classes, 1988.

- [CCITT Recommendation X.509|ISO 9594-8], [Recommendation for Directory|International Standard for Directory]: Authentication Framework, 1988.

- CCITT Recommendation F.435, Recommendation for Message Handling: EDI Messaging Service, forthcoming.

- CCITT Recommendation X.435, Recommendation for Message Handling Systems: EDI Messaging System, forthcoming.

- ISO 9735, Electronic data interchange for administration, commerce and transport (EDIFACT) - Application level syntax rules, 1987.

- The X.400 Blue Book Companion, CCITT X.400 MHS 1988, ISO/IEC MOTIS Message Oriented Text Interchange System, by Carl Uno Manros, published by Technology Appraisals, 1989.

- The EDI Handbook: Trading in the 1990s, edited by Mike Gifkins and David Hitchcock, published by Blenheim Online Publications, 1988.

- What is EDI?, by Martin Preston, published by NCC Publications, 1988.

- Electronic Data Interchange and Paperless Trading, The Implementation Guide, by Euromatica, published by Euromatica, 1988.

- L'EDI pour l'entreprise, by Victor Sandoval, published by Hermes, 1990

- La Technologie de l'EDI, by Victor Sandoval, published by Hermes, forthcoming.

Index

This index does not contain references to the Glossary, nor to Chapter 16 Road map.